New Perspectives on
Einstein's
$$E = MC^2$$

Other Related Titles from World Scientific

Group Theory in Physics: A Practitioner's Guide
by Rutwig Campoamor-Stursberg and Michel Rausch de Traubenberg
ISBN: 978-981-3273-60-3

Theoretical and Mathematical Physics: Problems and Solutions
by Willi-Hans Steeb
ISBN: 978-981-3275-37-9
ISBN: 978-981-3275-96-6 (pbk)

Theory of Groups and Symmetries: Finite Groups, Lie Groups, and Lie Algebras
by Alexey P Isaev and Valery A Rubakov
ISBN: 978-981-3236-85-1

Mathematics for Physics: An Illustrated Handbook
by Adam Marsh
ISBN: 978-981-3233-91-1

New Perspectives on
Einstein's

Young Suh Kim
University of Maryland, College Park, USA

Marilyn E Noz
New York University, USA

World Scientific

NEW JERSEY · LONDON · SINGAPORE · BEIJING · SHANGHAI · HONG KONG · TAIPEI · CHENNAI · TOKYO

Published by

World Scientific Publishing Co. Pte. Ltd.

5 Toh Tuck Link, Singapore 596224

USA office: 27 Warren Street, Suite 401-402, Hackensack, NJ 07601

UK office: 57 Shelton Street, Covent Garden, London WC2H 9HE

Library of Congress Cataloging-in-Publication Data

Names: Kim, Y. S., author. | Noz, Marilyn E., author.

Title: New perspectives on Einstein's E=mc² / Young Suh Kim (University of Maryland, College Park, USA), Marilyn E. Noz (New York University, USA).

Description: Singapore ; Hackensack, NJ : World Scientific Publishing Co. Pte. Ltd., [2018] | Includes bibliographical references and index.

Identifiers: LCCN 2018017442| ISBN 9789813237704 (hardcover ; alk. paper) | ISBN 9813237708 (hardcover ; alk. paper)

Subjects: LCSH: Space and time. | Lorentz transformations. | Special relativity (Physics)

Classification: LCC QC173.59.S65 K56 2018 | DDC 530.11--dc23

LC record available at https://lccn.loc.gov/2018017442

British Library Cataloguing-in-Publication Data

A catalogue record for this book is available from the British Library.

For any available supplementary material, please visit
https://www.worldscientific.com/worldscibooks/10.1142/10925#t=suppl

Desk Editor: Ng Kah Fee

Typeset by Stallion Press
Email: enquiries@stallionpress.com

Preface

In 1905, Albert Einstein formulated his special theory of relativity which led him to the formula $E = mc^2$. This formula states that the mass can be converted to energy. However, to physicists, this is the energy–momentum relation

$$E = \sqrt{(mc^2)^2 + (cp)^2},$$

applicable to particles with all possible speeds. This formula becomes

$$E = mc^2 + p^2/2m, \quad \text{and} \quad E = cp,$$

respectively, for slow and massive particles, and fast or massless particles. The slow massive particle has the energy of mc^2 in addition to the Newtonian kinetic energy of $p^2/2m$. Thus, every massive particle has a mass energy of mc^2. This is the discovery Einstein made in 1905 and is known as Einstein's $E = mc^2$.

In deriving this, he had to study the mathematics of Lorentz transformations, which conveys that space and time variables are linearly mixed for moving objects or moving observers. Things look differently for moving observers. The issue of how things look to observers with different speed is called *Lorentz covariance*. Indeed Einstein's energy–momentum relation is a Lorentz–covariant formula.

Niels Bohr was interested in why the energy levels of the hydrogen atom were discrete. Bohr's efforts led to the idea of wave–particle duality. The hydrogen atom energy levels are discrete because the electron orbits are standing waves. This eventually led to Schrödinger's wave mechanics in 1926 and Heisenberg's uncertainty principle in 1927.

These historical facts are well known. However, here is one question. Bohr and Einstein met occasionally to talk about physics. Bohr was interested in the hydrogen atom, while Einstein was worrying about how things look to moving observers.

While their interests are in two distinct directions, it was impossible for them not to talk about how the electron standing wave of the hydrogen atom looks to a moving observer. However, there are no records of their discussion on this subject. It is also possible that they never discussed this issue. At their time, it was not possible to think about the hydrogen atom moving with a relativistic speed. It was only a metaphysical problem.

The physics environment is quite different these days. High-energy accelerators routinely produce protons moving with speed very close to that of light. In addition, the proton is now a bound state of more fundamental particles known as the "quarks." It is believed that the quantum mechanics of binding the quarks is the same as the mechanics applicable to the hydrogen atom as an electron–proton bound state.

If Bohr and Einstein did not discuss the issue of moving standing waves, it is a good problem for us to study. The purpose of this book is based on the papers we published on this problem since 1973. We are not the first ones to notice this Bohr–Einstein problem.

Many distinguished physicists wrote many important papers on this fundamental question. Among them, we have been guided by the papers written by Paul A. M. Dirac (1902–1984), Eugene P. Wigner (1902–1995), and Richard P. Feynman (1918–1988).

This book is based on our efforts to unify the ideas of these three physicists to address the Bohr–Einstein gap. We construct in this book a model of the bound state in the Lorentz–covariant world. As Einstein's energy–momentum relation gives two different expressions for slow and fast particles, the standing wave for a bound state appears differently for slow and fast speeds.

It took Isaac Newton 20 years to extend his law of gravity for particles to spherical objects like the sun and earth. He had to develop a new mathematical method known today as integral calculus. In carrying out the program of extending Einstein's special relativity to extended objects, like the hydrogen atom or the proton in the quark model, we need a new mathematical tool, as illustrated in Fig. 1.

Fortunately, we did not have to invent this tool. It was already formulated by Eugene Paul Wigner in his 1939 article in the Annals of Mathematics, as is illustrated in Fig. 2. This fact is largely unknown. The purpose of this book is to give an interpretation of his work by giving physical examples.

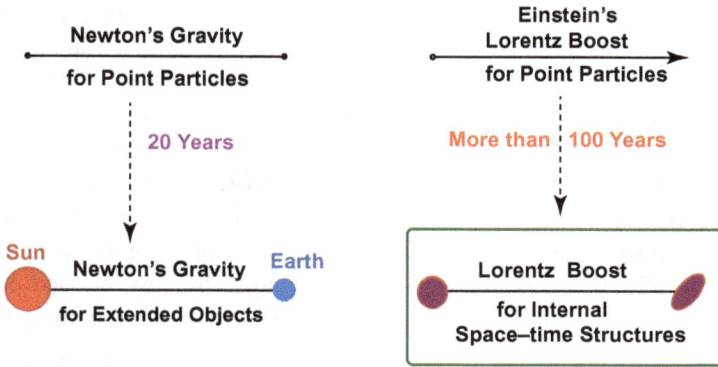

Fig. 1. Physical laws for point particles and for particles with internal space-time extensions. Einstein's law for Lorentz boosts is well known. Even more than 100 years after his formulation of special relativity in 1905, the issue of Lorentz covariance is not completely settled for internal space–time structures.

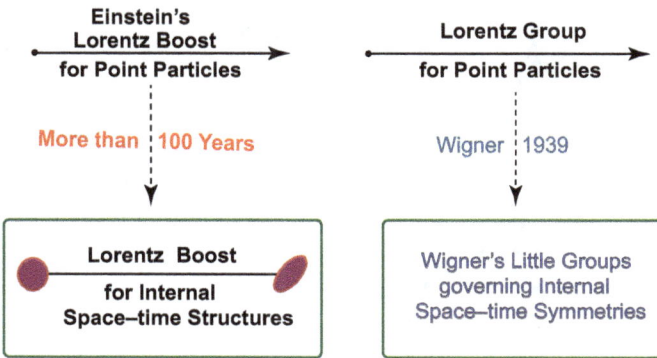

Fig. 2. The tool for internal space–time symmetries was formulated in 1939 by Eugene Wigner, but this fact is still largely unknown among the present generation of physicists.

Wigner's 1939 paper is known as the most difficult paper to read among the present generation of physicists. It is based on the Lorentz group which scares away young physicists these days. For this reason, we avoid group theory terminology as much as possible in this book and translate his paper into the language of 2×2 matrices.

We realize that Einstein's gravitational wave is of current interest. This phenomenon is based on Einstein's general relativity and is beyond the scope of this book. The point is that Einstein is still known for his $E = mc^2$, and this formula is based on his special theory of relativity formulated in 1905. By studying Wigner's 1939 paper, we explore further the contents of Einstein's $E = mc^2$ or his special relativity.

During the period of more than 40 years of our collaboration since 1973, we have benefited from new ideas contributed by our colleagues, namely, Michael Ruiz, Seog Oh, Thomas Karr, Daesoo Han, Dongchul Son, Sibel Başkal, and Elena Georgieva. One of us (YSK) was fortunate enough to visit Professor Eugene Wigner at Princeton University regularly during the period of 1985–1990. Needless to say, we could not have carried out our research program without Professor Wigner's advice and encouragement.

YSK and MEN

Contents

Chapter 1

Introduction

When Einstein formulated his special relativity in 1905, he worked out the transformation law for point particles. The question is what happens when those particles have space–time extensions. The hydrogen atom is a case in point. The hydrogen atom is small enough to be regarded as a particle obeying Einstein's law of Lorentz transformations including the energy–momentum relation $E = \sqrt{(mc^2)^2 + (cp)^2}$.

Yet, the hydrogen atom has a very rich internal space–time structure, rich enough to provide all the ingredients for quantum mechanics. Indeed, Niels Bohr was interested in why the energy levels of the hydrogen atom are discrete. His interest led to the replacement of the electron orbit by a standing wave.

Before and after 1927, Einstein and Bohr met occasionally to discuss physics. Einstein even stayed for an extended period at the Bohr Institute in Copenhagen, shown in Fig. 1.1.

It is possible that they discussed how the hydrogen atom with an electron orbit or as a standing wave looks to moving observers. However, there are no written records of their discussions. It is also possible that they never discussed this issue, since there were no hydrogen atoms moving with relativistic speeds, and/or since Einstein maintained a distance from the Copenhagen interpretation of quantum mechanics.

In either case, this is the issue to be resolved by the present generation of physicists. The present authors are not the first ones to raise this question. Among the many distinguished physicists who made laudable contributions to this fundamental problem, we should mention Paul A. M. Dirac (1902–1984) and Richard P. Feynman (1918–1988). They accepted the present

Fig. 1.1.   Niels Bohr Institute in Copenhagen. The plaque on the building says *This building was the place where the foundations of atomic and modern physics were laid in a scientifically creative environment inspired by Niels Bohr in the 1920s and 1930s.* The third floor of this building was used for hotel rooms for the visitors invited by Niels Bohr. Einstein was one of those visitors [Photos by Y. S. Kim (2015)].

form of quantum mechanics, and the Lorentz covariance on which Einstein's special relativity is based.

In this book, we translate what they did into the mathematical language of Wigner's little groups. Wigner in 1939 published a paper in the Annals of Mathematics (Wigner, 1939) discussing the subgroups of the Lorentz group whose transformations leave the momentum of a given particle invariant. These subgroups are called Wigner's little groups. It is noted that some additional work is needed for this book to achieve contact with Dirac's papers and Feynman's papers and to make contact with the real world of physics.

In Secs. 1.1–1.3, we explain how the papers by Dirac, Feynman, and Wigner serve useful purposes for this book. In Sec. 1.4, we explain the scope of contents of the book.

Throughout this book, the concept of *Lorentz covariance* and the mathematics of Lorentz transformations constitute the basic scientific language. They will be discussed in detail in Chapter 5.

1.1   Dirac's Approach

We are all familiar with the Dirac equation. This equation is for spin-1/2 particles in the relativistic world. Its success in describing the electron–positron is well known. It also describes the Thomas effect in the spin–orbit coupling. It is known that Dirac was able to write his equation purely from mathematical considerations. He believed in mathematical beauty.

Dirac was the first one interested in the uncertainty relation between time and energy (Dirac, 1927). He noted that unlike the position and momentum variables, there are no quantum excitations. This was one of the difficulties in making quantum mechanics covariant in the Lorentzian world where the space and time variables become linearly mixed for moving observers.

Yet, Dirac never gave up his love for mathematical beauty. In 1945 (Dirac, 1945b), Dirac attempted to construct representations of the Lorentz group using harmonic oscillator wave functions. He had to start with the Gaussian form

$$\exp\left\{-\frac{1}{2}\left(x^2 + y^2 + z^2 + (ct)^2\right)\right\}.$$

However, he did not explain how the time variable is defined. Does this function mean the world becomes zero in the remote past or remote future?

In his paper of 1949 (Dirac, 1949) entitled *Forms of Relativistic Dynamics*, Dirac presented three possible ways to make quantum mechanics Lorentz-covariant. He introduces many fresh ideas, such as the instant form, light-cone coordinate system, and the Poincaré group as an extension of Heisenberg's uncertainty relation. In this paper, Dirac mentions only difficulties, and makes no effort to provide resolutions.

In his paper of 1963 (Dirac, 1963), Dirac starts with two harmonic oscillators and exploits the symmetries contained in this dynamical system. He ends up with the symmetries of the group $O(3,2)$, namely, the Lorentz group with three spatial variables and two time variables. The harmonic oscillator is the language of quantum mechanics and the Lorentz group is the language of special relativity. Indeed, this paper tells us to construct a harmonic oscillator wave function which can be Lorentz-transformed.

1.2   Feynman's Approach

Richard Feynman made a giant step toward constructing Lorentz-covariant quantum mechanics. We are all familiar with Feynman diagrams. His diagrams serve as a tool for computing the scattering matrix or S-matrix applicable only to scattering processes. It was Feynman who realized this, and recommended harmonic oscillator wave functions for bound states in the Lorentz-covariant world (Feynman *et al.*, 1971).

Earlier, in 1969 (Feynman, 1969a,b), Feynamn observed that the proton, when it moves with a velocity very close to that of light, appears like a collection of an infinite number of partons, unlike the bound state in the quark model (Gell-Mann, 1964). The proton is regarded as a quantum bound state of three quarks when it is at rest. We assume that the quantum mechanics applicable to this bound state is the same as that applicable to the hydrogen atom as a bound state of the proton and electron.

In his 1972 book on *Statistical Mechanics* (Feynman, 1998), when he introduces the density matrix, Feynman says:

> *When we solve a quantum-mechanical problem, what we really do is divide the universe into two parts — the system in which we are interested and the rest of the universe. We then usually act as if the system in which we are interested comprised the entire universe. To motivate the use of density matrices, let us see what happens when we include the part of the universe outside the system.*

It is a challenge for us to find the rest of the universe in the Lorentz-covariant world. Is there a variable we do not measure in this Lorentz-covariant system?

Like Dirac, Feynman wrote many papers with the purpose of combining quantum mechanics with Einstein's special relativity. However, his style of presentation is quite different from that of Dirac. We have listed their differences in Table 1.1. On the other hand, this table shows a possibility of combining their ideas for constructing a Lorentz-covariant form of quantum mechanics.

Table 1.1   Dirac and Feynman.

	Paul A. M. Dirac	Richard P. Feynman
Means of expression	Poems	Diagrams cartoons
Personality	Seldom talks to anyone	Willing to join the crowd anywhere in the world
Primary interest	Standing waves Bound states	Running waves Scattering states
Favorite tools of research	Four-by-four matrices Harmonic oscillators	Path integrals Harmonic oscillators

Notes: Two different persons for the same purpose. They did not realize that their papers can be translated into the language of Wigner's little groups which dictate the internal space-time symmetry of particles in the Lorentz–covariant world.

1.3   Wigner's Little Groups

In this book, we choose to achieve this goal by translating Dirac's papers and Feynman's papers to the mathematics of Eugene Wigner which dictates the internal space–time symmetries of particles in the Lorentz-covariant world.

As for Wigner's mathematics, we are talking about his 1939 paper entitled *Representations of the Inhomogeneous Lorentz Group* published in Annals of Mathematics (Wigner, 1939). This paper also requires a translation. In this paper, Wigner constructed the subgroups of the Lorentz group whose transformations leave the momentum of a given particle invariant. These subgroups are called Wigner's little groups. Fortunately, these little groups can be translated into the language of 2×2 matrices and harmonic oscillators.

Indeed, using matrix algebra, it is possible to derive the result given in the second row of Table 1.2. Using the oscillator representation, it is possible to derive the result given in the third row of the same table. A particle at rest can have three spin orientations. If it is Lorentz-boosted, the longitudinal component remains as the helicity. The transverse components collapse into one gauge degree of freedom.

For bound states, we believe that the same quantum mechanics is applicable to the hydrogen atom and to the proton in the quark model. While the hydrogen atom cannot be accelerated, modern accelerators

Table 1.2   Further contents of Einstein's $E = mc^2$.

	Massive slow	Lorentz covariance	Fast massless
Energy Momentum	$mc^2 + p^2/2m$	Einstein's $E = \sqrt{(mc^2)^2 + (cp)^2}$	$E = cp$
Helicity spin, gauge	S_3 S_1, S_2	Wigner's Little Group	Helicity Gauge Trans.
Quarks in proton	Quark model	Covariant Oscillator	Parton Picture

Notes: The first row shows that Einstein's Lorentz-covariant expression gives two different forms of the energy–momentum relation for slow (massive) particles and fast (massless) particles. The second row illustrates that Wigner's little groups gives two different expressions for the massive and massless particles. The third row shows that Gell-Mann's quark model for slow hadrons and Feyman's parton model are two different limits of one covariant entity.

routinely produce protons moving with speed very close to that of light. For these fast-moving protons, Feynman, in 1969, formulated his parton picture (Feynman, 1969a,b).

In this picture, the proton consists of a collection of partons which appear to be quite different from the quarks. Indeed, using the oscillator representation of Wigner's little groups, it is possible to describe the quark model and the parton picture as two limiting cases of one Lorentz-covariant entity (Kim and Noz, 1977). This result is given in the third row of Table 1.2.

1.4   Scope of This Book

In Chapter 2, we shall discuss Einstein as a philosopher. What kind of philosopher was he? On his gravestone in London, Karl Marx says *The philosophers have only interpreted the world in various ways. The point however is to change it.* Einstein was most certainly a world-changing philosopher. Einstein started as a Kantianist (Howard, 2005), but he became a world-changing philosopher of his own kind. In Chapter 3, we give a broad outline of Einstein's life until he left Europe for the United States in 1933. In Chapter 4, we detail Einstein's life in the United States and his influence on the development of the nuclear bomb.

In Chapters 5–7, we introduce the Lorentz group as an algebra based on 2×2 matrices, and discuss Wigner's little groups applicable to internal space–time symmetries of particles in the Lorentz-covariant world. It is shown that the Dirac equation serves as a representation of the little groups.

In Chapter 8, we show it is possible to construct a representation of Wigner's little group for massive particles using harmonic oscillator wave functions. In Chapter 9, we use these Lorentz-covariant oscillators for some of the outstanding issues in high-energy physics. It is shown that the bound-state quark model for the slow proton and the parton picture of fast-moving proton are two limiting cases of one Lorentz-covariant entity. We thus complete Table 1.2.

In Chapter 10, it is noted that the density matrix allows us to deal with the variables which are not observed or observable in quantum mechanics. According to Feynman, they belong to the rest of the universe. In the covariant formulation, the time separation between two bound particle plays an important role, but this variable is in the rest of the universe. We study the effect of this variable when the bound state is Lorentz-boosted.

In Chapter 11, it is noted that harmonic oscillators and 2×2 matrices are applicable to all branches of physics. It is shown that the 2×2 representation of the Lorentz group and harmonic oscillators constitute the underlying scientific language for classical optics, quantum optics, and information theory.

Chapter 2

Einstein's Philosophical Base

Einstein is known as a physicist, but he is also known as a philosopher. If that is the case, what kind of philosopher was he? Karl Marx seems to know the answer to this question as seen in Fig. 2.1.

We all know how Einstein changed the world with his $E = mc^2$. The question is where Einstein stands among the philosophers who shaped our way of thinking.

It is known that Einstein's conceptual base for his theory of relativity was the philosophy formulated by Immanuel Kant (Howard, 2005). Things appear differently to observers in different frames. However, Kant's Ding-an-Sich leads to the existence of the absolute reference frame which is not acceptable in Einstein's theory. It is possible to avoid this conflict using the ancient Chinese philosophy of Taoism where two different views can co-exist in harmony.

This Taoist view is not enough to explain Einstein's formulation $E = mc^2$, which takes the form of

$$E = \sqrt{(mc^2)^2 + (cp)^2},\tag{2.1}$$

which becomes

$$E = mc^2 + \frac{p^2}{2m}, \quad \text{and} \quad E = cp,\tag{2.2}$$

in the limits of small and large p/mc, respectively.

By synthesizing the two formulae of Eq. (2.2) into the one formula of Eq. (2.1), Einstein followed the Hegelian way of approaching problems.

Isaac Newton synthesized open orbits for comets and closed orbits for planets to create his second law of motion. Maxwell combined electricity

on the marble wall at the entrance
lobby of Humboldt University in
Berlin.

on his grave stone at the Highgate ——→
Cemetery in London.

Fig. 2.1.   Karl Marx on philosophers. He says *The philosophers have only interpreted the world in various ways. The point however is to change it.* Marx is talking about Einstein yet to be born [Photos by Y.S. Kim (2008).]

and magnetism to create his four equations leading to the present-day wireless world. In order to reconcile wave and particle views of matter, Heisenberg formulated his uncertainty principle. Special relativity and quantum mechanics are the two greatest theories formulated in the 20th Century. It is the mission of the present generation of physicists to synthesize these two theories. For this purpose, let us examine Einstein's philosophical background.

2.1   Introduction

Einstein studied the philosophy of Immanuel Kant during his earlier years (Howard, 2005). It is thus not difficult to see he was influenced by the Kantian view of the world when he formulated his special theory of relativity. It is also known that, in formulating his philosophy, Kant was heavily influenced by the environment of Königsberg where he spent the entire 80 years of his life. The first question is what aspect of Kant's city was influential to Kant. We shall start with this issue in this chapter.

In Einstein's theory, one object looks differently to observers moving with different speeds. This aspect is quite consistent with Kant's philosophy. According to him, one given object or event can appear differently to observers in different environments or with different mindsets. In order to resolve this issue, Kant had to introduce the concept of *Ding-an-Sich* or *thing-in-itself* meaning an ultimate object of absolute truth. Indeed, Kant had a concept of relativity as Einstein did, but his Ding-an-Sich led to the absolute frame of reference. Here, Kantianism breaks down in Einstein's theory. Kant's absolute frame does not exist according to Einstein. In order to resolve this issue, let us go to the ancient Chinese philosophy of Taoism.

Here, there are two different observers with two opposite points of view. However, this world works when these two views form a harmony. Indeed, Einsteinism is more consistent with Taoism. The energy–momentum relation is different for a massive-slow particle and for a fast-massless particle. Einstein's relativity achieved the harmony between these two formulae. This leads us to Hegel's approach to the world. If there are two opposite things, it is possible to derive a new thing from them. This is what Einsteinism is all about. Einstein derived his $E = mc^2$ from two different expressions of the energy–momentum relation for massive and massless particles. Einstein thus started with Kantianism, but he developed a Hegelian approach to physical problems. Indeed, this encourages us to see how this Hegelianism played the role in developing new laws of physics. For instance, Newton's equation of motion combines the open orbits for comets and the closed orbits of planets.

If this Hegelian approach is so natural to the history of physics, there is a good reason. Hegel derived his philosophy by studying history. Hegel observed that Christianity is a product of Jewish one-God religion and Greek philosophy. Since Hegel did not understand physics, his reasoning was based on historical development of human relations. It is thus an interesting proposition to interpret Hegel's philosophy using the precise science of Physics.

In Secs. 2.2 and 2.3, we review how Kantianism was developed and how Einstein was influenced by Kant. In Sec. 2.4, it is pointed out that Hegelianism is the natural language for creating a new physics based on two established theories. In Sec. 2.5, we observe that Einstein plays a transitional role from Kant to Hegel. In Sec. 2.6, it is noted that Max Born developed his reciprocity principle from the symmetry between the space–time and energy–momentum coordinates. Finally, in Sec. 2.7, we examine whether Einstein was interested in hearing about Taoism from Hideki Yukawa who was a Taoist scholar (Tanikawa, 1979).

2.2 Geographic Origin of Kantianism and Taoism

Immanuel Kant (1724–1804) was born in the East Prussian city of Königsberg, and there he spent the 80 years of his entire life. It is agreed that his philosophy was influenced by the lifestyle of Königsberg. The city of Königberg (now Kaliningrad) is located at the Baltic wedge between Poland and Lithuania. It is also between two large lagoons. This place served as the traffic center for maritime traders in the Baltic Sea. In addition, this city is between the eastern and western worlds. However, there are no natural boundaries such as rivers or mountains. Thus, anyone with a stronger army could come to this area and run the place (Applebaum, 2017).

For instance, Königsberg was a German city when Kant wrote his books. The same city became a Russian city of Kaliningrad after the Soviet troops occupied that area during World War II. Immanuel Kant is still respected there, and the university there is called *Kant State University*. The city maintains a museum dedicated to Kant as shown in Fig. 2.2.

Indeed, Königsberg was a meeting place for many people with different ideas and different view points. Kant observed that the same thing can appear differently depending on the observer's location or state of mind.

The basic ingredients of Taoism are known to be two opposite elements Yang (plus) and Ying (minus). This world works best if these two elements form a harmony. However, the most interesting aspect of Taoism is that its geographic origin is the same as that of Kantianism.

After the ice age, China started as a collection of isolated pockets of population. They then came to the banks of the Yellow River, and started to communicate with those from other areas. They drew pictures for written communication leading eventually to Chinese characters. How about different ideas? They grouped many different opinions into two

Fig. 2.2.   Kant museum and Kant's grave in Kaliningrad, Russia. Kaliningrad was the German city of Königsberg when Kant was born and lived there for 80 years. Since Kant did not believe in Jesus, his grave is outside the Königsberg Cathedral. Russians respect Kant, and the museum and the grave are well kept. [Photos by Y.S. Kim (2005).]

groups, leading to the concept of Yang and Ying. Immanuel Kant considered many different views, but he concluded that there must be one and only one truth. Indeed, Taoism and Kantianism started with the same environment, but Kant insisted on one truth called Ding-an-Sich, while Taoism ended up with two opposing elements. It is however interesting to note that both Kantianism and Taoism were developed from the same geographical condition, namely different people coming to one place.

We can understand how Taoism was developed by looking at a more recent historical development. The North American continent was a new land for European immigrants. Those Europeans wanted to create a government which would reflect all different opinions. The result was to create two political parties.

2.3   Kantian Influence on Einstein

During his early years, Einstein became quite interested in Kant and studied his philosophy rigorously (Howard, 2005). This was quite common among young students during his time. Einstein however studied also physics and got an idea that one object could appear differently for observers moving with different speeds. Let us go to Fig. 2.3. According to Kant, an object or event looks differently to different observers depending on their places or states of mind. A Coca-Cola can looks like a circle if viewed from the top. It appears like a rectangle if viewed from the side. The Coca-Cola can is an absolute thing or his Ding-an-Sich. Likewise, the electron orbit of the

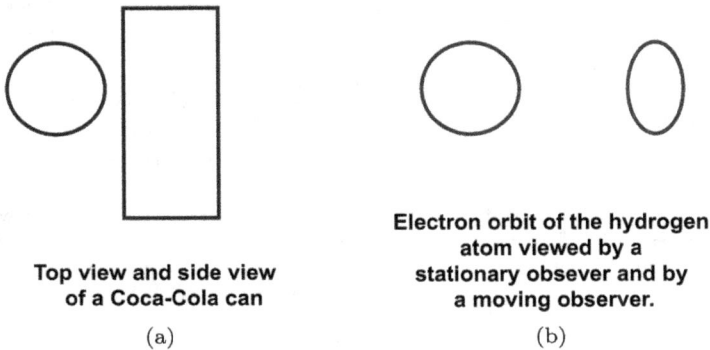

**Top view and side view
of a Coca-Cola can**

(a)

**Electron orbit of the hydrogen
atom viewed by a
stationary obsever and by
a moving observer.**

(b)

Fig. 2.3.   A Coca-Cola can appears differently to two observers from two different angles. Likewise, the electron orbit in the hydrogen atom should appear differently to two observers moving with two different speeds (Bell, 2004).

hydrogen atom looks like a circle for an observer when both the hydrogen atom and the observer are stationary. If the hydrogen atom is on a train, our first guess is that it should look like an ellipse. This is what Einstein inherited from Kant.

Let us come back to Einstein. Like Kant, Einstein started from different observers looking at a thing differently, but ended up with a particle at rest and the same particle moving with a speed close to that of light. He then derived his celebrated energy–mass relation, as indicated in Fig. 2.4. Einstein had to invent a formula applicable to both. This is precisely a Hegelian approach to physics. It is not clear whether Einstein knew he was doing Hegel. This remains as an interesting historical problem. As for Einstein's hydrogen atom, we now have hadrons which are bound states of quarks, while the hydrogen atom is a bound state of a proton and an electron. The proton is a hadron and is a charged particle which can be accelerated to the speed very close to that of light.

Yes, Einstein was interested in how things appear to two different observers: one stationary and the other moving with velocity v along the z direction. Let their coordinates be z and z' respectively. Then their coordinates are related by

$$z' = z + vt \qquad (2.3)$$

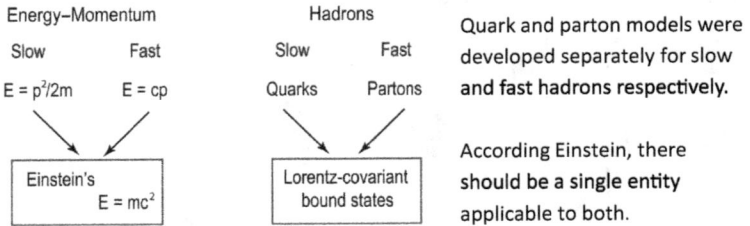

Fig. 2.4.   The energy–momentum relation of a particle takes a different form when the particle moves with different speeds. Let us choose two limiting cases. Einstein was able to find the same formula applicable to both. In so doing, he found his $E = mc^2$ which means $E = \sqrt{(mc^2)^2 + (cp)^2}$. Likewise, the quark model (Gell-Mann, 1964) and the parton model (Feynman, 1969a,b) should produce a Lorentz-covariant picture of the bound state. Does this mean that the system requires an absolute Lorentz frame, or Kant's Ding-an-Sich? The answer is No. Indeed, Kant attempted to formulate his theory of relativity with an absolute coordinate system corresponding to his Ding-an-Sich. This is the basic departure of Einsteinism from Kantianism.

in the traditional world. However, by the time of Einstein, it was known that the time variable also undergoes a change, so that

$$t' = t + \frac{vz}{c^2}, \tag{2.4}$$

where c is the speed of light. In our daily life, the velocity of a moving object, such as a train or airplane is so small compared with c, we can ignore the change in the t variable. However, the situation is quite different when v is comparable with c as in the case of protons coming from high-energy accelerators.

Indeed, Eqs. (2.3) and (2.4) lead to

$$(z')^2 - (ct')^2 = \left[1 - \left(\frac{v}{c}\right)^2\right](z^2 - c^2 t^2), \tag{2.5}$$

and to the transformation law

$$\begin{pmatrix} z' \\ ct' \end{pmatrix} = \begin{pmatrix} \cosh\eta & \sinh\eta \\ \sinh\eta & \cosh\eta \end{pmatrix} \begin{pmatrix} z \\ ct \end{pmatrix}, \tag{2.6}$$

with

$$\sinh\eta = \frac{v}{\sqrt{c^2 - v^2}}, \quad \cosh\eta = \frac{c}{\sqrt{c^2 - v^2}}, \tag{2.7}$$

which can be combined into

$$\frac{v}{c} = \tanh\eta. \tag{2.8}$$

The transformation given in Eq. (2.6) is known as the Lorentz transformation. The translation of physics from one value of η to that of another value of η is called the *Lorentz covariance*. The equality of Eq. (2.5) leads to

$$(ct')^2 - (z')^2 = (ct)^2 - (z)^2, \tag{2.9}$$

Fig. 2.5. Hegelhaus in Stuttgart (Germany) and a portrait of Hegel talking to his publisher. [Photos by Y.S. Kim (2012).]

which is independent of parameter η. If the quantities remain invariant under Lorentz transformations, they are called *Lorentz invariants*.

Like Kant, Einstein was interested in observers in different environments. He was interested in physics with different values of η. He was particularly interested in combining the physics with $\eta = 0$ (at rest) with the physics with $\eta \to 1$ (moving as fast as light).

2.4 Hegelian Approach to the History of Physics

Since Hegel formulated his philosophy while studying history, it is quite natural to write the history of physics according to Hegel. First of all, Isaac Newton combined hyperbolic-like orbits for comets and elliptic orbits for planets to derive his second-order differential equation which is known today as Newton's second law, as is illustrated in Fig. 2.6.

James Clerk Maxwell combined the theory of electricity and that for magnetism to formulate his electromagnetic theory leading to the present world of wireless communication. Einstein noted Newton's mechanics and Maxwell's equations obey different transformation laws. He made both to obey the same transformation law, leading to his theory of relativity. Max

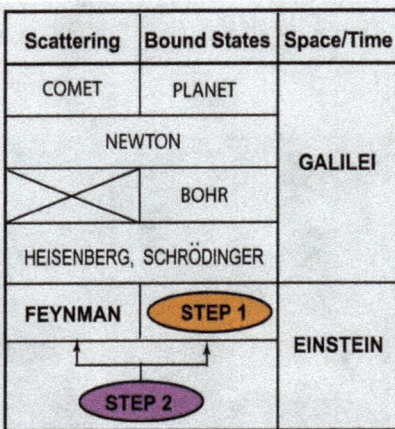

Scattering	Bound States	Space/Time
COMET	PLANET	
NEWTON		GALILEI
	BOHR	
HEISENBERG, SCHRÖDINGER		
FEYNMAN	STEP 1	EINSTEIN
STEP 2		

Humans developed the theories of electricity of magnetism separately. James Clerk Maxwell combined them into one theory, leading to the present form of wireless world. However, the transformation law of his theory is different from that of Newtonian mechanics. This was another challenge to Einstein.

Tranformation	Newton	Maxwell
Galilei	Yes	No
Lorentz	No	Yes

Fig. 2.6.   Progress in physical theories is made when two theories are combined into one. The left side of this figure shows how mechanics was developed. The right side shows how mechanics and electromagnetism were led to obey the same transformation law.

Planck observed that the radiation laws are different for low- and high-frequency limits. By deriving one formula for both, he discovered Planck's constant.

Werner Heisenberg observed that matter appears as a particle and also appears as a wave, with entirely different properties. He found the common ground for both. In doing so, he found the uncertainty relation which constitutes the foundation of quantum mechanics. Indeed, quantum mechanics and special relativity were the two most fundamental theories formulated in the 20th century. They were developed independently. The question is whether they can be combined into one theory. We shall examine how the Hegelian approach is appropriate for this problem in Sec. 2.5.

2.5 Einstein between Kant and Hegel

Einstein started as Kantianist. However, it is clear that he became a Hegelianist while formulating his special relativity as shown in Fig. 2.4. Thus, he plays a transitional role from Kant and Hegel.

In order to emphasize his role as a Hegelianist, let us consider another way in which Einstein derived his $E = mc^2$. By the time of Einstein, the rule of Lorentz transformations was well established for the space and time coordinates, thanks to Lorentz, Poincaré, and Minkowski.

This undoubtedly led Einstein to suspect the rule of Lorentz covariance also applied to momentum space. The Lorentz transformation applicable to the Minkowski space of (x, y, z, t) was well known to him. For the space coordinates of (x, y, z), their conjugate momentum coordinates are (p_x, p_y, p_z). Einstein's idea for the variable conjugate to the time variable was E, the energy of the particle. Einstein thus considered a conjugate Minkowski space of

$$(p_x, p_y, p_z, E/c), \tag{2.10}$$

to which the rule of Lorentz transformations is applicable.

In the Lorentz-covariant world, the quantity

$$(ct)^2 - x^2 - y^2 - z^2 \tag{2.11}$$

remains invariant under transformations. Likewise, the quantity

$$E^2 - p_x^2 - p_y^2 - p_z^2. \tag{2.12}$$

should remain constant. According to Einstein, this quantity is mc^2. This becomes

$$E = \sqrt{(mc^2)^2 + (cp)^2}, \tag{2.13}$$

with

$$p^2 = p_x^2 + p_y^2 + p_z^2. \tag{2.14}$$

Einstein's logic presented in this section is not a straight-forward application of Hegelianism. However, this logic was influenced by Hegel or could be called *Reverse-Hegelianism*.

2.6 Born's Reciprocity Principle

Inspired by this symmetry, Max Born came up with an idea of relativistic dynamics based on the symmetry between space–time and momentum–energy. Born's effort along this direction is known as the reciprocity principle (Born, 1938, 1949). His key point is whether the phase space can be made Lorentz-covariant.

Inspired by Born's reciprocity, Hideki Yukawa in 1953 noted that the equation of motion for the harmonic oscillator takes the same form for both momentum and space coordinates. In addition, the space–time and momentum–energy coordinates are not Lorentz-covariant. With this observation, he constructed harmonic oscillator wave functions that can be Lorentz-transformed (Yukawa, 1953).

Indeed, the authors of this book wrote their first joint paper based on the Lorentz-covariant oscillator wave function written down by Yukawa in his 1953 paper (Yukawa, 1953).

2.7 Einstein and Yukawa

While in the United States, from 1933 to 1955, Einstein did not talk to too many physicists. Hideki Yukawa was one of those small number of people with whom he talked to. In 1948, Hideki Yukawa was invited to Princeton's Institute for Advanced Study and stayed there for 1 year until he moved to Columbia University in 1949.

The question is what they talked about. In view of the paper Yukawa published while at Princeton (Yukawa, 1949), it is possible that they talked about Born's reciprocity principle. However, Einstein could have been more interested in hearing about Taoism from Yukawa. Yukawa was known as a Taoist scholar who was capable of reading the original Chinese books written by Laotze and Changtze more than 2500 years ago (Tanikawa, 1979). In view of Einstein's keen interest in Kantianism, it is possible that he wanted to hear about Taoist philosophy from Yukawa.

Einstein, Yukawa, and Wheeler (1953). Credit: Johns Hopkins University.

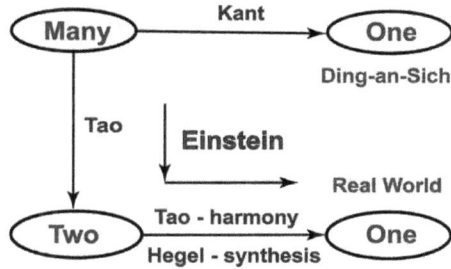

Fig. 2.7. Map of philosophers. This map tells how Einstein's philosophical background is working in deriving his $E = mc^2$. He started with $E = p^2/2m$ for all massive particles, and $E = cp$ for all massless particles. Einstein then synthesized these two formulae to $E = \sqrt{m^2c^4 + (cp)^2}$. This tells us why Einstein was interested in talking with Yukawa (in the middle) who was a Taoist. The other person (in white dress) is John A. Wheeler who was interested in converting Einstein's general relativity into experimentally testable science. There was also a good reason for Einstein to be happy with Wheeler.

In Taoism, many things or events converge into two contrasting groups. The harmony of these two groups makes this world go round. However, Taoism does not tell how to synthesize these two groups into one. Einstein synthesized two different formulae into one as described in Fig. 2.4. Thus, there is every reason to believe that Einstein was interested in where he stands in the map of philosophers.

Based on the reasoning given in this chapter, we can construct a map of the philosophers as shown in Fig. 2.7. Philosophers construct their opinions of the world from their observations of social behaviors and historical developments, because they do not understand physics. In terms of physics, Einstein gives a precise way in which the philosophers can develop their theories.

Chapter 3

More about Einstein

Einstein was born in Ulm in 1879, but his family moved to Munich when he was an infant. In Munich, Einstein's father, together with his younger brother, owned and operated a very profitable electric appliance factory with up to 200 employees. In Munich, the Einstein family enjoyed an affluent life.

Alas in 1894, their company became bankrupt. While their factory was producing DC-based electrical appliances, the AC revolution put those DC factories out of business. Thus, when Einstein was 15 years old, he had to go to Italy with his parents. They first moved to Milan and then to Pavia.

One year later, in 1895, at the age of 16, Albert went to Switzerland. There, he finished his high-school education. Two years later in 1897, Einstein entered Zürich Federal Polytechnic Institute for his 4 years of college program. This school is known today as ETH (Eidgenössische Technische Hochschule).

Like all physicists, Einstein had to spend frustrating years of job hunting after his graduation. In 1903, he found a position and married Mileva Marić who was his classmate at the ETH. He became a Research Assistant at the Swiss patent office in Bern, Switzerland and stayed there until 1909. After numerous scientific achievements, Einstein went back to the ETH in Zurich as an Associate Professor.

While in Bern, in 1905, Einstein completed his history-making work on special relativity leading to his $E = mc^2$. In the same year, he was awarded a PhD degree by the University of Zurich. Einstein was 26 years old then.

After spending 1 year (1911–1912) as a Professor at Charles University in Prague, Einstein returned to the ETH as a Professor. There, with Marcel Grossmann, Einstein studied Riemannian geometry dealing with curved space–time. Einstein and Grossmann were classmates when they were students at the ETH. This is how Einstein became deeply interested in developing his general theory of relativity. In 1911, Einstein concluded that a light ray coming to earth from a star should be bent by the sun's gravity. In 1916, Einstein completed his theory of general relativity.

In 1914, Einstein moved to Berlin to become a Professor at Humboldt University. While in Berlin in 1919, after a prolonged separation and divorce from his first wife Mileva, Einstein married Elsa Lowenthal. It is said that Einstein had a romantic relation with Elsa for many years before their marriage. Elsa came to the United States with Einstein in 1933, but she died in 1936.

Hitler's persecution of Jews forced Einstein to immigrate as a refugee to the United States in 1933. Einstein had his house in Princeton and his office at the Institute for Advanced Study. He became a citizen of the United States in 1940.

One year earlier, in 1939, Einstein signed a letter to President Roosevelt urging him to develop nuclear bombs. This is the reason why Einstein is known to the public as a bomb builder, but this is not the exact description of Einstein's role in developing nuclear bombs.

In Sec. 3.1, we point out that Einstein developed his special theory of relativity while working for the Swiss patent office. The city of Bern is quite proud of being the birth place of Einstein's $E = mc^2$ and maintains two museums dedicated to him.

In Sec. 3.2, it is pointed out that, while in Bern, Einstein did the ground work for quantum mechanics. He published papers on the photoelectric effect and the specific heat of solids. In these papers, Einstein gave a physical meaning to Planck's constant.

In 1921, Einstein received the Nobel Prize in physics for his work on the photoelectric effect, but not for his relativity theories. In Sec. 3.3, we list the direct experimental consequences of his special relativity and $E = mc^2$.

3.1   Einstein's Bern, the Birth Place of $E = mc^2$

Einstein spent 7 years in Bern, Switzerland (1903–1909) while working for the Swiss patent office. While in Bern, Einstein lived with his wife in an apartment building at 49 Kramgasse. During this period, Einstein completed his special theory of relativity in 1905. In 2005, in commemoration of 100th year of Einstein's formulation of relativity, the city of Bern converted this building to a museum called Einsteinhaus, as can be seen in Fig. 3.1.

In this house, during the years 1903-1905, Albert Einstein created his fundamental treatise on relativity theory.

Fig. 3.1. Einsteinhaus in Bern. With his first wife, Einstein lived on the 2nd floor of this apartment building. In front of this building, there is brass plate saying Einstein completed the theory of relativity during the period 1903–1905. [Photos by Kim (1997)].

In front of this building, there is brass plate saying Einstein formulated his relativity theory at this place.

Since this museum is dedicated to Einstein, there are numerous photos and displays. It is assumed that Einstein met many physicists and philosophers while in Bern. Indeed, Einstein spent his most creative years while living in this building. This museum contains many interesting photos, writings, and illustrations about Einstein. The following items are particularly interesting:

- Photo of Einstein's parents.
- Photo of Einstein with his first wife.
- Einstein's Genealogy, as complicated as other notable persons.
- Einstein's passport pages. Before coming to the United States, Einstein traveled from Germany to many different countries including Italy, Switzerland, Czechoslovakia, and Japan.
- The posters giving a brief description of Einstein's life in Bern (1903–1909), and description of his life in Berlin.
- A photo of the Swiss patent office where Einstein worked.
- Photos showing Einstein a music lover. He played a violin.
- The desk and chair he used when he was writing his articles on relativity.
- Photos of Einstein's life at ETH before coming to Bern.

Einsteinhaus　　　　　　　　**Einstein Museum**

Fig. 3.2.　Two museums in Bern dedicated to Einstein. In 2005, in order to celebrate 100 years of Einstein's $E = mc^2$, the city of Bern converted the apartment building where Einstein lived into a museum called Einsteinhaus. In the same year, the city created the Einstein Museum within its history museum complex. [Photos by Kim (2012)].

In 2005, in order to celebrate the centennial year of $E = mc^2$, the city of Bern created another museum within the Historical Museum of Bern complex as shown in Fig. 3.2. This Einstein Museum is located about 2 km south of the Einsteinhaus (within walking distance). The Einstein Section was refurbished to meet modern high-tech standards.

This museum contains a vast collection of Einstein-related materials, including many photos, magazine covers, numerous awards including his Nobel Prize certificates, as well as anti-semitic materials aimed against him.

There are photos of the electric appliance factory the Einstein family once owned in Munich. According to the belongings shown in this museum, Einstein came from a very affluent family. The following items particularly show how rich his family was in Munich.

- Photos of the electric appliance factory owned and operated by the Einstein family.
- A replica of the family dinner table while in Munich.
- The typewriter and telephone owned by the family.
- One of his class photos when Einstein went to his school in Munich.

However, this company became bankrupt in 1894, and Einstein's family had to move to Italy. One year later, in 1895, Einstein came to Switzerland to continue his studies. After finishing his secondary education, Einstein entered ETH for college education. The museum contains some photos of his life while he was a student there. The museum shows off many photos of his life in Bern and Berlin, and in the United States, including the following items:

- There is a photo of Einstein working in the patent office. He looks quite hopeless.
- Einstein's Nobel Prize citation. He received his Nobel Prize in 1921 for his interpretation of the photoelectric effect, not for his formulation of the theory of relativity.
- Photos of Einstein with his second wife, Elsa Lowenthal.
- Cartoon of Einstein as a Jewish animal being kicked out from Europe.
- Time Magazine covers carrying Einstein's images, including the image of Einstein with that of Marilyn Monroe, as two of the most popular Americans during the period 1950–1955.
- The luggage cases he used when he immigrated to the United States in 1933, indicating that Einstein was not a poor person.

Yes, the city of Bern maintains these two excellent museums to show the life of Albert Einstein. However, these museums do not show anything about his philosophical inclinations. For instance, they could show how Einstein became interested in Immanuel Kant during his childhood.

3.2 Einstein's Contribution to Quantum World

Einstein made important contributions to quantum physics during its early years, with his interpretation of the photoelectric effect and his calculation of the specific heat of solids. Einstein was able to make these contributions only because Max Planck introduced a phenomenological constant when he constructed his formula for black-body radiation.

Planck was born in 1858 and died in 1947. He spent his final years in the city of Göttingen which served as the headquarters of European physics during the 1930s. He weathered the war years of World War II, but it appears that he was too old to play a role in Hitler's project to develop nuclear bombs.

(a)　　　　　　　　　　　　　　　(b)

Fig. 3.3.　Statues of Max Planck (1858–1947) (a) and Hermann von Helmholtz (1821–1894) (b) in the front yard of Humboldt University in Berlin. Helmholtz was also a history–making physicist. He formulated the law of energy conservation, among others. [Photos by Kim (2017)].

Planck was a Professor at Humboldt University of Berlin from 1889 until he moved to Göttingen in 1928. His statue, as a humble-looking professor, is in the front yard of this university as shown in Fig. 3.3.

Planck observed that black-body radiation density, according to Rayleigh and Jeans, is

$$B_T(\omega) = \left(\frac{kT}{8\pi^4} \right) \omega^4, \tag{3.1}$$

where k is Boltzmann's constant. This formula is valid only for small values of ω. On the other hand, according to Wien's observation, the upper limit on this distribution is

$$\hbar\omega_{\max} = (0.4965)kT. \tag{3.2}$$

In 1900, Planck concluded that the radiation density should be (Planck, 1900)

$$B_T(\omega) = \left(\frac{\hbar\omega^3}{4\pi^3 c^2} \right) \frac{1}{e^{(\hbar\omega/kT)} - 1}. \tag{3.3}$$

This formula indeed gives the radiation density for all values of the frequency ω. This is widely known as Planck's radiation formula. From this formula, Planck concluded that $\hbar\omega$ is a measure of energy.

Table 3.1   Frogs and birds in physics. In deriving his radiation formula, Planck not only discovered a constant named after him, but also introduced two different ways of looking at physics. As we emphasized in Chapter 2, this is a Hegelian approach to physics, which Einstein later used deriving his energy–momentum relation.

	Frogs	Birds	Frogs
Black-body radiation	Rayleigh–Jeans	Planck	Wien
Energy–momentum	$p^2/2m$	Einstein's $E = \sqrt{(cp)^2 + (mc^2)^2}$	$E = cp$

When Planck derived this formula, the frequency was measured in ν, and $\hbar$ in h, where $\omega = 2\pi\nu$ and $\hbar = h/2\pi$. Thus, in his original papers, he used the notation $h\nu$ for $\hbar\omega$. The constant h was an unknown constant which converts the frequency ν into the units of energy and became known later as Planck's constant. In this book, for convenience, we shall call $\hbar$ and ω Planck's constant and frequency, respectively.

In addition to the introduction of this new constant, Planck introduced a new way of looking at physics. He developed a *bird's view* and a *frog's view* toward physics. Let us go to Table 3.1. Frogs developed the local laws, and birds synthesized the local laws into one global law. This approach to physics was discussed in Chapter 2 as a Hegelian process.

Yes, $\hbar\omega$ is measured in units of energy, but what kind of energy is it? Einstein's interpretation of the photoelectric effect answers this question.

If a metal plate is heated, electrons come out with speed. Likewise, if a light beam hits the metal plate, electrons also come out. This was known before Einstein became interested in this subject. Einstein noted that the kinetic energy of emitted electrons satisfy the relation

$$\text{Kinetic energy} = \hbar\omega - \phi, \tag{3.4}$$

where ω is the frequency of the incoming light, and ϕ was a constant depending on materials.

This means that each electron absorbs the energy of $\hbar\omega$ and uses ϕ to overcome the potential barrier to come out of the metal. This means that the light wave can be considered as a collection of particles carrying the energy of $\hbar\omega$.

Einstein with his $E = \sqrt{(cp)^2 + (mc^2)^2}$ was able to interpret that the light wave can be regarded as a collection of massless particles whose energy is pc which can be converted to $\hbar\omega$. These particles then have a momentum equal to $\hbar\omega/c$. By doing this, Einstein introduced the concept of massless particles called *photons*. In this way, Einstein defined the role of Planck's constant as a tool for defining the dual nature of matter as particles and waves.

Einstein had to face a stiff resistance from his colleagues on this issue. How can waves be particles? However, he received strong support from Planck. Max Planck received his Nobel Prize in 1918, and Einstein got his in 1921 for his interpretation of photoelectric effect.

Einstein made another history-making contribution to quantum theory, namely discrete energy levels. In 1907, assuming that the energy of electrons take discrete values of $n\hbar\omega$, Einstein calculated that the heat capacity of solids decreased rapidly as the temperature decreases.

Later in 1926, Erwin Schrödinger, using his wave mechanics, produced equal-spaced energy levels for the harmonic oscillator. This is the reason why Einstein's specific heat is called the harmonic oscillator model. It is remarkable that Einstein constructed his formula for heat capacity without Schrödinger's picture of quantum mechanics.

Einstein's formula gives the rapid decrease, but does not give an accurate description near $T = 0$. In order to fix this problem, Peter Debye had to introduce the concept of phonons in 1912. This aspect is well known.

Yet, we all know how essential the concept of discrete values of energy was to later development of wave mechanics and quantum mechanics. Einstein introduced this concept. By 1927, Einstein became a prominent figure in the community of physicists, as shown in the photo of the participants of the 1927 Solvay Conference held in Brussels (Fig. 3.4).

3.3 Experimental Verification of Einstein's Special Relativity

Einstein did not get the Nobel prize for relativity because the Nobel committee members did not believe his general relativity was a complete theory. They thought his special relativity serves only as a stepping stone toward the general theory. The Nobel Committee could not fully appreciate the physical consequences predicted by Einstein's special relativity, which came in later years. Let us discuss some of the early experimental consequences.

In 1930, Wolfgang Pauli noted there was an inconsistency in the decay of the neutron into a proton and an election, not only in terms of the conservation of angular momentum, but also in terms of energy and momentum conservation. In order to preserve these indispensable

Fig. 3.4. Participants of the Solvay Conference 1927. From back to front and from left to right: Auguste Piccard, Émile Henriot, Paul Ehrenfest, Édouard Herzen, Théophile de Donder, Erwin Schrödinger, Jules-Émile Verschaffelt, Wolfgang Pauli, Werner Heisenberg, Ralph Howard Fowler, Léon Brillouin, Peter Debye, Martin Knudsen, William Lawrence Bragg, Hendrik Anthony Kramers, Paul Dirac, Arthur Compton, Louis de Broglie, Max Born, Niels Bohr, Irving Langmuir, Max Planck, Marie Sklodowska Curie, Hendrik Lorentz, Albert Einstein, Paul Langevin, Charles-Eugéne Guye, Charles Thomson Rees Wilson, Owen Willans Richardson. (Photo from the public domain).

conservation laws as well as Einstein's energy–momentum relation, Pauli then predicted a new particle called a neutrino, as shown in Fig. 3.5(a). The neutrino is a neutral particle and was not detectable with technologies available at his time.

In 1935, Hideki Yukawa predicted a particle heavier than an electron and lighter than a proton (Yukawa, 1935). This new particle was called the meson. Using the cloud chamber he used to detect positrons in 1932 (Anderson, 1933), Carl Anderson in 1936 discovered mesons in cosmic rays bombarding this earth. They were initially understood as the mesons predicted by Yukawa in 1935. They were later determined to be different, and were called μ mesons or muons. The lifetime of this meson is approximately 2 micro-seconds.

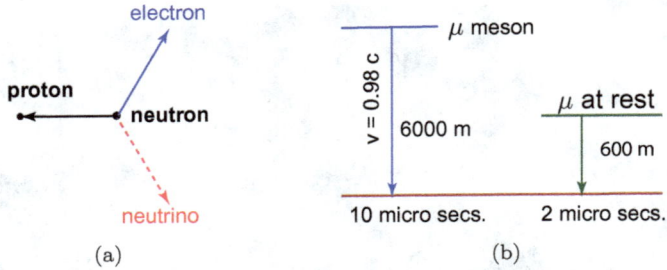

electron

proton

neutron

neutrino

(a)

μ meson

$v = 0.98\,c$

6000 m

μ at rest

600 m

10 micro secs.

2 micro secs.

(b)

Fig. 3.5. Experimental proof of Einstein's special relativity. Neutron decaying into the proton, election, and neutrino. There is a mass difference of 2.4 MeV. Thus, during the decay process, this energy is transferred to the electron mass, electron kinetic energy and the energy of the massless neutrino, while the total momentum is conserved. The μ meson is shown as decaying at different rates depending on whether it is traveling with a speed close to that of light or is at rest.

In Fig. 3.5(b), a muon is created 6000 m above the ground and decays when it comes down to the ground after 10 micro-seconds. The speed of the muon is 0.98 c. On the other hand, measured from the muon rest frame, it reaches the ground after 2 micro-seconds. This phenomenon clearly establishes the validity of Einstein's special relativity which includes time dilation and space contraction.

Thus far, all the particles with relativistic speeds were from cosmic rays. In 1929, an American physicist named Ernest Lawrence came up with an idea of accelerating a proton by a machine, called the cyclotron. His first cyclotron was constructed in 1931. The idea is to let the proton circle around in the constant magnetic field as shown in Fig. 3.6. The frequency of this circular motion and the time period required to make one complete motion are

$$\omega = \frac{eB}{mc}, \qquad T = \frac{2\pi mc}{eB},$$

respectively, where m and B are the proton mass and the magnetic field, respectively. These formulae do not depend on the radius of the circular motion. Thus, the proton with a larger radius moves faster.

According to Einstein's special relativity, the period becomes

$$T = \frac{2\pi mc}{eB\sqrt{1 - (v/c)^2}}.$$

Thus, T should be modulated when the radius of the circle becomes larger.

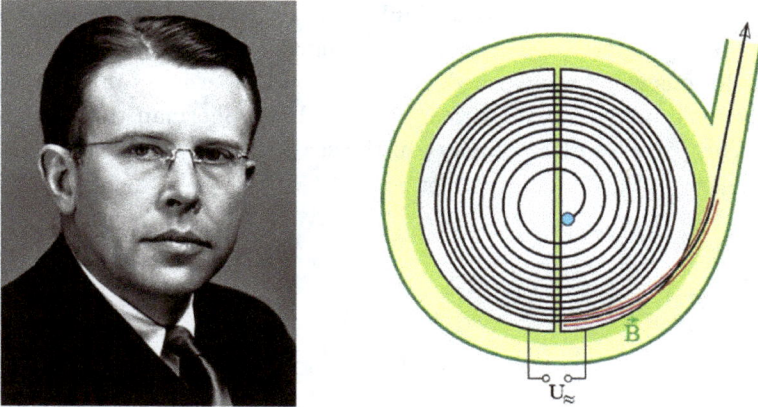

Fig. 3.6. Ernest Lawrence and expanding proton orbit in the cyclotron. The frequency of one cyclic motion is independent of speed. Thus, the proton with a larger radius moves faster. However, if the proton speed becomes comparable with that of light, the period of one cyclic motion becomes longer. This is due to Einstein's special relativity. (Photos from the public domain).

The particle accelerator taking into account this effect is called synchrocyclotron or frequency-modulated cyclotron. There were four synchrocyclotrons built in the United States by 1954, and this marked the beginning of high-energy particle physics.

Finally, let us discuss $E = mc^2$ with the numbers closer to our daily life. Unlike Roman Emperors, humans these days enjoy conveniences provided by electric power. It is possible to estimate the electric energy consumed by all humans for 1 year and convert it to mass. We can begin with the monthly electric energy consumption from the electric bills. The energy there is measured in kWh (kilowatt hour). We should convert 1 kWh to kilograms of mass first and go through the following steps:

1. 1 kWh is 3.6×10^6 joules. Thus, according to $m = E/c^2$, 1 kWh $= 4 \times 10^{-10}$ kg.
2. The average American household consumes 600 kWh of electric energy per month. Thus, 7200 kWh per year or 3×10^{-6} kg.
3. There are about 100 million households in the United States. Thus, 300 kg per year.

If we assume that the total yearly energy consumption by the entire world (including other forms of energy) is about 100 times of the above figure, it becomes 30 metric tons.

During World War II, Soviet-built T-34 tanks destroyed Hitler's army in the bitter battles of Kursk and Stalingrad. The mass of this tank is about 30 tons when fully armed. American-built Sherman tanks played the decisive role in the North African battles. Each Sherman tank weighs 33 tons. This gives an idea of energy in the scale of mass, thanks to Einstein.

Chapter 4

Einstein in the United States

Einstein made his first trip to the United States in 1921. When he arrived at New York's South Street Seaport (near Wall Street), Dean Henry Burchard Fine of Princeton University went there to greet him. He gave a series of five lectures at the University on his new theory of relativity. Scientists from all over the United States packed the lecture hall to listen to his lectures.

In 1933, Einstein immigrated to the United States as a refugee and became a U.S. citizen in 1940. He died at Princeton Hospital in 1955. While in the United States, he was a great public figure strong enough to persuade the President of the United States to develop the nuclear bomb.

In 1933, Einstein settled down in his house at 112 Mercer Street in Princeton. His office was at the Institute for Advanced Study within walking distance from his house.

He became known to every American after World War II. Americans thought that their first nuclear bomb was invented by Einstein. These days, he is known for mc^2 even to those who do not know the meaning of m or c.

In Sec. 4.1, we shall talk about his life in the town of Princeton, New Jersey. Section 4.2 shows photographs of visitors to his house not commonly seen.

Washington, DC is the capital city of the United States, with statues of great Americans. In Sec. 4.3, we shall tell where Einstein's statue is. The U.S. Naval Academy is in Annapolis, Maryland, 50 km east of Washington. Albert Michelson and Joseph Weber studied there. Michelson conducted the Michelson–Morley experiment, and Weber is regarded as the father of detection of Einstein's gravitational wave.

The most controversial Einstein issue is his role in building the first nuclear bomb. Yes, in 1939, he signed a letter to President Roosevelt giving his support for developing the bomb. In Sec. 4.4, we reprint the letter Einstein signed as well as Roosevelt's reply. We also discuss how high-energy nuclear-particle physics was developed after World War II.

Another controversial issue was and still is Einstein's refusal to accept the Copenhagen interpretation of quantum mechanics. In 1954, Werner Heisenberg visited Einstein to hear directly from him about the status of physics at that time. Heisenberg then wrote about this meeting as a lecture note. In Sec. 4.5, we give a review of this lecture note.

4.1 Einstein in Princeton

In the 1920s, Louis Bamberger and Caroline Bamberger Fuld operated a department store in Newark, New Jersey. They invested their earnings in the stock market, but they pulled out their investment in 1929 right before the market crash. They then decided to do something good for the state of New Jersey and in 1930, the Bambergers provided the founding gift of five million dollars to establish an institute in Princeton. This institute, known as the Institute for Advanced Study, was dedicated to the vision of Abraham Flexner who was an education reformer. It is reasonable to assume that Caroline was Bamberger's daughter, and her husband's last name was Fuld. The Institute's main building is called Fuld Hall.

It was Abraham Flexner, the first director of the Institute, who initiated the idea of inviting Albert Einstein from Germany. Einstein thus came to Princeton in 1933 where he lived and worked until his death in 1955.

The distance between his house on Mercer Street and his office at the Institute is about 2 km. Einstein used to walk to his office along the route shown in Fig. 4.1.

Einstein did not talk to too many American physicists while in Princeton. He interacted closely with the philosopher named Kurt Gödel who joined the Institute for Advanced Study in 1940, after immigration from Austria. His philosophy is beyond the scope of this book.

Einstein did not like the Copenhagen interpretation of quantum mechanics, even though this has been the accepted interpretation since 1927. It is possible that Einstein was interested in giving his own interpretation of quantum mechanics based on Gödel's philosophy while making quantum mechanics consistent with his theories of relativity. A photo of Einstein with Gödel is shown in Fig. 4.2. They had their offices at Fuld Hall of the Institute shown in the same figure.

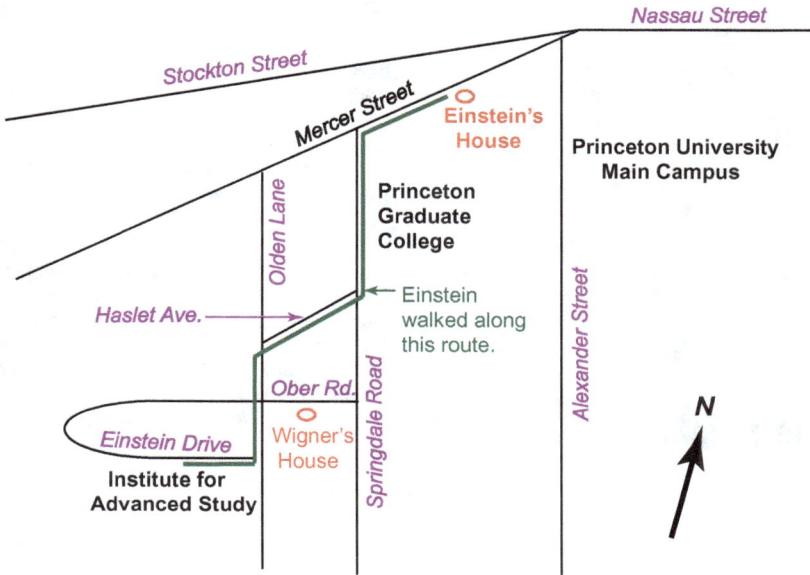

Fig. 4.1. Einstein route in Princeton. Einstein walked along the route specified in this map, to go to his office at the Institute for Advanced Study and to come back to his home at 112 Mercer Street.

Fig. 4.2. Gödel and Einstein (left), and the Fuld Hall of the Institute for Advanced Study in Princeton. (Gödel–Einstein photo from the AIP Emilio Segré Visual Archives, and photo of the Institute for Advanced Study by Y. S. Kim (2008).)

REVIEWS OF MODERN PHYSICS Vol. 21, Issue 3 (1949).

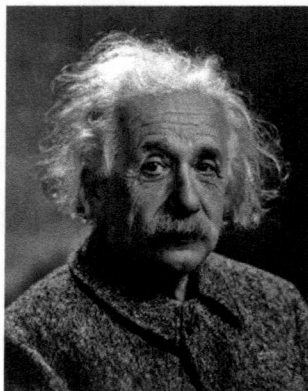

Authors: Robert A. Millikan, L. de Broglie, M. von Laue, Philipp Frank, M. S. Vallarta, Georges Lemaître, G. Gamow, R.C. Tolman, H.P. Robertson, S. Chandrasekhar, A.H. Taub, P.A.M. Dirac, T.D. Newton and E.P. Wigner, A. D. Fokker, L. Infeld and A. Schild, E.G. Straus, J.A. Schouten, John A. Wheeler and Richard P. Feynman, W. Pauli and F. Villars, A. Pais and S. T. Epstein, Kurt Gödel, H.J. Bhabha, Max Born, Hideki Yukawa, Peter G. Bergmann and Johanna H.M. Brunings, V. Bargmann, Giulio Racah, Cornelius Lanczos, Nathan Rosen, James Franck and Robert Livingston, R. Ladenburg and D. Bershader, Theodore von Kármán and C. C. Lin, Hans Albert Einstein and El-Sayed Ahmed El-Samni, W.J. de Haas and G.J. van den Berg, Karl F. Herzfeld, O. Klein, Otto Stern, Banesh Hoffmann.

Fig. 4.3.   Photo of Einstein and the list of authors in the special issue of the *Reviews of Modern Physics*. The photo of Einstein was produced by O. J. Turner of Princeton in 1947, who donated his copyright to the U.S. Library of Congress. It is thus in the public domain.

In 1949, Einstein became 70 years old. The American Physical Society was nice enough to publish a special issue of the *Reviews of Modern Physics* with articles contributed by the most distinguished physicists throughout the world, with Einstein's photograph taken in 1947. The list of authors in this special issue as well as the photo of Einstein is shown in Fig. 4.3.

This photo of Einstein was produced by Orren Jack Turner (1919–2008) of Princeton. He was born in Princeton and died there. As a professional photographer, he took photos of many Princeton events and persons.

There are many important papers in this issue of the *Reviews of Modern Physics*. Of particular importance is Dirac's article entitled *Forms of Relativistic Dynamics*. In this paper, Dirac discussed possible ways of making quantum mechanics consistent with Einstein's special relativity (Dirac, 1949). This paper is discussed in detail by (Kim and Noz, 1986).

4.2   Einstein's House

While Einstein's house in Princeton can be spotted easily, there are not many photos of Einstein with his house in the background. Figure 4.4 shows

Fig. 4.4.   Einstein's house in Princeton (Photo by Y.S. Kim, 2002), and Einstein talking with Irene Joliot-Curie on the front steps of his house. (Photo from John K. Webb, School of Physics, Univ. of New South Wales, Sydney, Australia.)

Einstein talking with Irene Joliot-Curie on the front steps of his house. This photo is owned by Prof. John K. Webb, School of Physics, University of New South Wales in Sydney, Australia. He found this photo at an antique shop in Paris. He thinks this photo was produced in 1945.

Irene's mother was of course Marie Sklodowska Curie. Irene continued her mother's research on radium and received in 1936 her Nobel Prize in chemistry.

Marie Curie was born in 1867, 12 years earlier than Einstein in 1879. She received the Nobel Prizes in physics (1903) and in chemistry (1911). She was thus the central figure at the first Solvay Conference held in 1911, while Einstein was a first-year full professor at Charles University in Prague. The photo of the 1911 Solvay Conference participants contains the images of both Curie and Einstein, as shown in Fig. 4.5.

J. Robert Oppenheimer went to Einstein's house. As the director of the Institute for Advanced Study, he had to check whether everything was OK there. In 1954, Werner Heisenberg visited his house, and his visit is discussed in detail in Sec. 4.5.

The town of Princeton is known for the university carrying the same name. Princeton University used to be an all-male school until 1969. Largely unknown is that there was and still is a predominantly female college in the same town, called Westminster Choir College. This college had and still has an excellent choir.

Every year, in front of Einstein's house, this choir used to sing in the morning of Christmas Day before the sun rises. They wanted to tell

Fig. 4.5. Participants of the First Solvay Conference in 1911. Marie Sklodowska Curie and Albert Einstein were among the participants. Seated (L–R): W. Nernst, M. Brillouin, E. Solvay, H. Lorentz, E. Warburg, J. Perrin, W. Wien, M. Curie, and H. Poincaré. Standing (L–R): R. Goldschmidt, M. Planck, H. Rubens, A. Sommerfeld, F. Lindemann, M. de Broglie, M. Knudsen, F. Hasenöhrl, G. Hostelet, E. Herzen, J.H. Jeans, E. Rutherford, H. Kamerlingh Onnes, A. Einstein, and P. Langevin. (Photo from the public domain.)

Einstein that Christ has come. Figure 4.6 shows ladies who were students of this college during the 1940s. They sang for Einstein. According to them, Einstein used to come out from his house to greet them.

The husbands of the two ladies in Fig. 4.6 are Richard Underwood (left) and Bill Holden, respectively. Richard's grandfather was Horace Underwood who was the first Presbyterian missionary to Korea. He had an elder brother named John Underwood. John was one of the pioneers of the typewriter industry in the United States. Underwood typewriters were once the dominant desktop machines as personal computers are today. The other gentleman was Bill Holden. He was a musician and was one of the band members who played the National Anthem during Einstein's naturalization ceremony in 1940. They seem to have their own Einstein connections, and interesting stories to tell.

According to an article published by Steven Schultz in the Princeton Weekly Bulletin in 2004 (Schultz, 2004), Einstein met frequently a Princeton librarian named Johanna Fantova. They met in 1930 while

Fig. 4.6. Two ladies who were students at Westminster Choir College during Einstein's time in Princeton. The students of this college used to sing Christmas carols in front of Einstein's house in the dawn of the Christmas Day. [Photos by Y.S. Kim (left 2003) and (right 2004).]

Fig. 4.7. Einstein with Johanna Fantova, at Lake Carnage, Princeton. [Photo from Princeton weekly Bulletin (April 26, 2004), released to the public domain.] Princeton students enjoying their boat ride at the same lake. [Photo by Y.S. Kim (1983).]

they both were in Europe. Fantova was with Einstein whenever he had important guests in his house. Figure 4.7 shows Einstein's happy moment with Fantova.

Einstein died at the Princeton Medical Center. After his death, the U.S. Internal Revenue Service (tax-collecting agency) disclosed that Einstein used to make mistakes in calculating his income taxes. However, the service was nice enough not to raise the issue.

4.3 Einstein in Washington

Einstein is widely respected all over the United States. He appeared very often on the cover of Time magazine. According to this magazine, Marilyn Monroe and Albert Einstein were the two most popular figures in the period 1950–1955.

The capital city of the United States is Washington, DC. It has many statues of those who made America great: George Washington, Thomas Jefferson, Abraham Lincoln, Andrew Jackson, Marquis de Lafayette, Ulysses S. Grant, Casimir Pulaski, William Sherman, Tadeusz Kosciuszko, Alexander Hamilton, Franklin Roosevelt, Martin Luther King, among others. There is a giant statue of Albert Einstein within walking distance from the White House as shown in Fig. 4.8. He is holding a book entitled $E = mc^2$. This monument is at the front of the headquarters of the National Academy of Sciences.

Many tourists come to see the statue of Einstein. To most of them who do not know physics, Einstein used to be a scientific genius who invented the nuclear bomb to finish World War II in 1945. These days, he is known as an American genius who invented $E = mc^2$, even though many do not understand the meaning of the letters E, m, and c.

The U.S. Naval Academy is in the town of Annapolis, Maryland, not far from the capital city of Washington. This college produced many talented naval officers capable of operating nuclear powered submarines and aircraft carriers. The Naval Academy is traditionally a high-tech engineering

Fig. 4.8.   Bronze statue of Albert Einstein and the White House in Washington, DC, U.S.A. The Einstein statue is 2 km west of the White House, and about 1 km east of the Lincoln Memorial. [Photos by Y.S. Kim (2013).]

college that produced two physicists whose names cannot be separated from Einstein. They were Albert Michelson and Joseph Weber. They were in the graduating classes of 1873 and 1940, respectively. They went to the Naval Academy because their parents could not afford their college expenses.

Albert Abraham Michelson was born in Poland in 1852 and came to the United States with his parents when he was 2 years old. When it was time for college, he took the entrance examination at the U.S. Naval Academy, and scored the top grade. Yet, he had difficulty in entering the school because of his Jewish background. He was admitted only after the courageous decision made by President Ulysses S. Grant.

Michelson developed precision devices to measure the velocity of light and atomic spectral lines. His obsession with the speed of light is well known. He also invented the atomic clock. He received the Nobel Prize in physics in 1907, and became the first American scientist to receive this honor. There is a shiny building dedicated to his name at the Naval Academy in Annapolis as shown in Fig. 4.9.

He learned about Einstein's theory of relativity, but he was dead set to find the absolute inertial frame, resulting in a series of negative results. These negative results of course greatly enhanced Einstein's theory of relativity.

Joseph Weber was born and raised in Patterson, New Jersey. He was interested in electronics from his childhood. In 1935, he entered Cooper

Albert A. Michelson and Michelson Hall at the U.S. Naval Academy in Annapolis.

Fig. 4.9. Albert Abraham Michelson in his navy uniform and Michelson Hall at the US Naval Academy. Photo of Michelson from the public domain. [Photo of Michelson Hall by Y.S. Kim (2015).]

Fig. 4.10. Joseph Weber in his navy uniform (Photo from the public domain), and his gravitational wave antenna (Photo from AIP Emilio Segré Visual Archives).

Union College for his undergraduate education. Even though Cooper Union has no tuition, his parents could not afford his room and board. Thus, he decided to go the Naval Academy in 1936 and was admitted after passing the entrance examination. Before making the final decision, the admission officer visited his house in Patterson. Weber used to say that the officer came to check whether he was an African–American because his home address was in a very poor area of New Jersey.

Weber was stationed at Pearl Harbor when Japanese planes attacked on December 7, 1941, and he performed many dangerous missions during World War II. In 1948, he joined the faculty of the University of Maryland (College Park) at the department of electrical engineering, while working for his PhD degree at Catholic University in Washington, DC. He received his degree in 1951. He later joined the physics faculty at the University of Maryland.

While in the navy, Weber was interested in microwave devices. His interest in this subject led to the idea of producing coherent light beams from simultaneous atomic transitions, and he submitted a paper in 1951 for the 1952 Electron Tube Research Conference held in Ottawa. This paper is known to be the earliest paper on the principles behind the laser and the maser. Based on Weber's idea, Charles Townes, then a Professor at Columbia University of the City of New York, and Nikolay Basov and Aleksandr Prokhorov of the P.N. Lebedev Physical Institute, Moscow, USSR produced maser and laser beams in their laboratories, and received their Nobel Prizes in 1964. (Note that in 1967, Townes was appointed University Professor at large by the University of California, based at Berkeley; he spent the rest of his long life at Berkeley.) Weber was jointly nominated for the prize, but he was left out. This is still a controversial issue.

During his sabbatical year 1955–1956, Weber met John A. Wheeler of Princeton. He then became interested in gravitational waves predicted by Einstein's general theory of relativity and started building gravitational antennae at the University of Maryland. However, the technology available at that time was not accurate enough for him to measure the gravitational wave.

In February 2016, the announcement was made that the gravitational wave was detected by the younger generation of physicists. Yet, it is agreed that they could not have achieved this result without Weber's determined efforts. Weber is regarded as the father of gravitational waves.

4.4   Einstein and the First Nuclear Bomb

To men or women on the street, Einstein is known as the man who built the first nuclear bomb dropped on the Japanese city of Hiroshima on August 6, 1945. This was not the case.

During World War II, Hitler's Nazi Germany attempted to construct bombs based on nuclear reactions. Afraid of Hitler's bomb, a group of Hungarian-born physicists, including Leo Szilard, Eugene Wigner, John von Neumann, and Edward Teller, advanced the idea of building an American bomb first. These Hungarian–Americans studied in Germany before coming to the United States, and they were quite familiar with the capability of German scientists and engineers in dealing with nuclear technology.

However, in order to build the bomb based on nuclear reactions in the United States, they needed a huge amount of support from the government. They needed money, security, mobilization of scientific facilities, and of all capable American scientists and engineers. The only way to get this support was to convince the president of the United States.

They decided to write a letter to President Roosevelt. The question then was whether their letter could reach the president's attention. The only available channel was to get Einstein to sign it. The president had to read Einstein's letters.

The letter was written by Leo Szilard. Szilard then went to Einstein with Eugene Wigner and Edward Teller and asked him to sign the letter. The final version was in English, but they drafted the letter in German first. They talked about the danger of a possible Hitler's bomb. Einstein, who was vacationing on Long Island, NY, signed the letter. Einstein later said he signed only because he was afraid of what Hitler could do.

Einstein signed the following letter to Roosevelt.

_____ **Einstein's letter to Roosevelt**

> Albert Einstein
> Old Grove Rd.
> Nassau Point
> Peconic, Long Island

August 2nd, 1939
F. D. Roosevelt,
President of the United States,
White House
Washington, D.C.

Sir:

Some recent work by E. Fermi and L. Szilard, which has been communicated to me in manuscript, leads me to expect that the element uranium may be turned into a new and important source of energy in the immediate future. Certain aspects of the situation which has arisen seem to call for watchfulness and, if necessary, quick action on the part of the Administration. I believe therefore that it is my duty to bring to your attention the following facts and recommendations:

In the course of the last four months it has been made probable — through the work of Joliot in France as well as Fermi and Szilard in America — that it may become possible to set up a nuclear chain reaction in a large mass of uranium, by which vast amounts of power and large quantities of new radium-like elements would be generated. Now it appears almost certain that this could be achieved in the immediate future.

This new phenomenon would also lead to the construction of bombs, and it is conceivable — but much less certain — that extremely powerful bombs of a new type may thus be constructed. A single bomb of this type, carried by boat and exploded in a port, might very well destroy the whole port together with some of the surrounding territory. However, such bombs might very well prove to be too heavy for transportation by air.

The United States has only very poor ores of uranium in moderate quantities. There are some good ores in Canada and the former Czechoslovakia, while the most important source of uranium is Belgian Congo.

In view of the situation you may think it desirable to have some permanent contact maintained between the administration and the group of physicists working on chain reactions in America. One possible way of achieving this might be for you to entrust with this task a person who has your confidence and who could perhaps serve in an unofficial capacity. His task might comprise the following:

a. to approach Government Departments, keep them informed of the further development, and put forward recommendations for government action, giving particular attention to the problem of securing a supply of uranium ore for the United States;
b. to speed up the experimental work, which is at present being carried on within the limits of the budgets of University laboratories, by providing funds, if such funds be required, through his contacts with private persons who are willing to make contributions for this cause, and perhaps also by obtaining the cooperation of industrial laboratories which have the necessary equipment.

I understand that Germany has actually stopped the sale of uranium from the Czechoslovakian mines which she has taken over. That she should have taken such early action might perhaps be understood on the ground that the son of the German UnderSecretary of state, von Weiszäcker, is attached to the Kaiser-Wilhelm-Institut in Berlin where some of the American work on uranium is now being repeated.

Yours very truly,

Albert Einstein

THE WHITE HOUSE
Washington

October 19, 1939

My dear Professor:

I want to thank you for your recent letter and the most interesting and important enclosure.

I found this data of such import that I have convened a Board consisting of the head of Bureau of Standards and a chosen representative of the Army and Navy to thoroughly investigate the possibilities of your suggestion regarding the element of uranium.

I am glad to say that Dr. Sachs will cooperate and work with this Committee and I feel this is the most practical and effective method of dealing with this subject.

Please accept my sincere thanks.

Very sincerely yours,

(signed) Franklin D. Roosevelt

Dr. Albert Einstein,
Old Grove Road,
Nassau Point,
Peconic, Long Island,
New York.

_____

    Roosevelt's reply marks the beginning of the massive secret government program to build nuclear bombs, known as the Manhattan project. This project required the mobilization and organization of all nuclear physicists in the United States and Western Europe. Niels Bohr came to one of the bomb-making centers in New Mexico, now called *Los Alamos National*

Fig. 4.11. Harry Truman signing the bill (1946) creating the Atomic Energy Commission for peaceful use of nuclear energy, and one of the first four post-war synchrocyclotrons built for Carnegie Institute of Technology in Pittsburgh, Pennsylvania. [Photos from the public domain.]

Laboratory. He was called Nicholas Baker. Richard Feynman and Roy Glauber were young men then.

The question is what happened after the war. In 1946, the U.S. government decided to convert this scientific establishment to peaceful use of nuclear energy, and created an agency called the *Atomic Energy Commission,* as shown in Fig. 4.11. The first project in this peaceful program was to build particle accelerators which can produce protons moving with relativistic speed. Again the Einstein issue.

The Atomic Energy Commission initially built four synchrocyclotrons producing protons with the kinetic energy of 450 MeV or with the speed of 0.75c. One of them was at the Nuclear Research Center of Carnegie Institute of Technology (now called Carnegie Mellon University). This laboratory was located at Saxonburg, Pennsylvania about 50 km north of the main campus. One of the authors of this book (YSK) worked at this laboratory during the summer of 1957. He was an undergraduate student then.

After the first wave of cyclotrons, the U.S. government kept building more powerful and expensive particle accelerators, creating the golden era of high-energy physics. This led to Murray Gell-Mann's quark model (Gell-Mann, 1964) for the proton at rest and Feynman's parton picture (Feynman, 1969a,b). The question is whether these two different views are consistent with Einstein's special relativity. This is one of the major issues we would like to address in this book.

After building so many accelerators in the United States, the U.S. government decided not to fund new accelerators. Now the largest accelerator is at the Large Hadron Collider Center near Geneva, Switzerland supported by participating countries. Needless to say, all these expensive scientific activities started from Einstein's $E = mc^2$ of 1905.

4.5 Heisenberg on Einstein

This section is a review of Heisenberg's essay entitled *Encounters and Conversations with Albert Einstein.* This essay is based on Heisenberg's lecture delivered at the Einsteinhaus in Ulm (Germany) on June 27, 1974 (see Fig. 4.12). This lecture note was translated into English and was included in a book published by Princeton University Press (Heisenberg, 1989). This book consists of nine essays Heisenberg wrote on people, places, and particles. The book was copyrighted by Werner Heisenberg in 1983, presumably by the Heisenberg estate. Its English version was published by Seabury Press (San Francisco) in 1983 as *Tradition in Science,* and was reprinted by Princeton University Press in the Princeton Science Library Edition by arrangement with Harper and Row in 1989.

Werner Heisenberg was born in 1901 and died in 1976. He was 4 years old when Einstein formulated special relativity in 1905. Ten years later, when he was in high school, Heisenberg became interested in Einstein's theory and started his physics career out of his respect for Einstein. However,

Fig. 4.12. Einstein Monument at his birthplace and an Einstein artwork in front of the museum called *Einsteinhaus* in Ulm, Germany. [Photos by Y.S. Kim (2004).]

these two great physicists did not like each other. What went wrong? The basic point is well known. Einstein never accepted Heisenberg's uncertainty principle as a fundamental physical law.

Let us see how Heisenberg became a physicist.

• Heisenberg liked mathematics and became interested in special relativity when he was very young. The mathematics of Lorentz transformations was easy for him to understand, but the physical concept of simultaneity was very difficult for him to grasp. This probably was the communication gap he had with Einstein. When he was in college in Munich, he learned about Einstein and his theories from Arnold Sommerfeld who was a great teacher to him. Sommerfeld also recognized Heisenberg's potential and encouraged him to meet Einstein personally. The first step toward this process was to attend Einstein's lectures.

• In addition, Heisenberg became quite interested in atomic physics which was Sommerfeld's main subject. He was interested in the question of why classical theories fail to explain atomic phenomena, and how the concept of light quanta, formulated by Einstein, could explain those *anomalies*. As is well known, Heisenberg's concentrated effort to resolve those puzzles led him to formulate his uncertainty principle in 1927. In the above-mentioned book, which contains Heisenberg's article about Einstein, he has chapters entitled *Development of Concepts in the History of Quantum Mechanics,* and *The Beginnings of Quantum Mechanics in Göttingen.* Quite understandably, they constitute the first and second chapters of his book.

In the summer of 1922, the Society of German Scientists and Physicists had a meeting in Leipzig, and Einstein was scheduled to give a lecture. Sommerfeld encouraged Heisenberg to attend Einstein's talk. When he went there, a young man gave him a red leaflet saying that the theory of relativity is a totally unproved Jewish speculation, and that it had been undeservedly amplified through Jewish newspapers on behalf of Einstein, a fellow-member of their race. Heisenberg noted there that those leaflets were being handed out by the students of Germany's most respected experimental physicist at that time. Heisenberg did not mention his name, but it is not difficult to know who that most respected experimentalist was. Instead of Einstein, von Laue gave his lecture. Heisenberg's first attempt to meet Einstein failed in this way.

In early 1926, Heisenberg was invited to give a colloquium on his quantum mechanics by the physicists in Berlin. At that time, Berlin was the citadel of physics. The audience included Planck, von Laue, Nernst, and Einstein. Einstein was quite interested in Heisenberg's talk, and invited Heisenberg to come to his house. This was his first meeting with Einstein. However, Einstein was not happy with Heisenberg's interpretation of his new mechanics. Einstein's position was that every theory in fact contains unobservable quantities. The principle of employing only observable quantities simply cannot be consistently carried out.

In the spring of 1927, Heisenberg succeeded in formulating his uncertainty relation, emboldened by the wave mechanics formulated by Erwin Schrödinger in 1926 where electrons are regarded as waves. In the autumn of 1927, Heisenberg met Einstein again at the Solvay Conference in Brussels. Both Einstein and Heisenberg are in the photo shown in Fig. 3.4. There, Einstein came up with one counter-example to the uncertainty relation each morning, but, by dinner time, Heisenberg together with Bohr and Pauli were able to prove that Einstein's example was consistent with the uncertainty principle.

Three years later in 1930, Heisenberg met Einstein again at another Solvay Conference in Brussels. There, Bohr did his best to convince Einstein that the uncertainty relation is a fundamental law in physics. Einstein still refused, and they agreed to disagree. It was Heisenberg's last time to see Einstein in Europe. By 1933, the political environment became much worse in Germany, and Einstein moved to the United States. He lived and worked in Princeton where he had given his earlier lectures in 1921.

In 1954, Heisenberg visited Einstein's house in Princeton. Heisenberg was warned by Einstein's assistant not to stay with him more than one hour because of his poor health. Yet, Heisenberg recollects that Einstein was kind enough to spend almost the entire afternoon with him. They talked only about physics, but Einstein's position on the uncertainty relation remained unchanged. Again, Heisenberg failed to get Einstein's endorsement of his principle as a fundamental physical law. Einstein died in 1955, and Heisenberg died in 1976. According to Johanna Fantova, who was Einstein's close friend in Princeton, Einstein was not happy with Heisenberg when they met in 1954 (Schultz, 2004).

Needless to say, Heisenberg visited Einstein in order to make him happy. Thus, before going to Princeton in 1954, Heisenberg could have done the following homework.

1. He could have studied the paper published in 1935 by Einstein, Podolsky, and Rosen which is widely known as the EPR paper these days (Einstein *et al.*, 1935).

2. Heisenberg pointed out that Einstein once declared himself as a pacifist. He then said, in view of his support for the development of nuclear weapons in the United States, Einstein was not an absolute pacifist, but a pacifist with some adjective. Here, Heisenberg forgot to mention whether Hitler's Nazi set-up was an absolute evil or an evil with a different adjective.

3. In his article, Heisenberg says the concept of Einstein's simultaneity was very difficult to digest. That is right, the simultaneity (or simultaneous measurements) plays a pivotal role in interpreting the uncertainty relation. Yet, Heisenberg's interpretation of his uncertainty principle does not take into account the Lorentz covariance dictated by special relativity. If Heisenberg had studied this covariance question in interpreting his uncertainty principle, he could have drawn more interest from Einstein.

4. Paul A. M. Dirac had written many papers on the issue of making the uncertainty relation consistent with special relativity. Heisenberg could have talked with Dirac on this issue before making his trip to Princeton.

The EPR problem has been one of the fundamental questions discussed by many physicists even these days. However, Einstein could have been happier if Heisenberg had considered Einstein's Lorentz covariance when he formulated his uncertainty principle. This is the major issue we would like to address in this book. Einstein formulated his quantum concept while studying the Lorentz-covariant energy–momentum relation or $E = mc^2$ for massive and massless particles. Thanks to this relation, the photon is a massless particle.

Chapter 5

Introduction to the Lorentz Group

We are quite familiar with rotations and the representations of the rotation group. We are also familiar with the Lorentz boost in one direction. If these two operations are combined, the result is the Lorentz group. Both the rotation and Lorentz groups are continuous groups, which are called *Lie groups*. For these groups, it is convenient to use their generators. For the three-dimensional rotation group, there are three generators, and they form a closed set of commutation relations. This closed set is called the *Lie algebra*.

If we augment the rotation group with the Lorentz boost, the result is the Lorentz group. In Sec. 5.1, we introduce the group of 4×4 matrices applicable to the four-dimensional Minkowskian space consisting of three space dimensions and one time dimension. There are six generators for this group and its Lie algebra consists of a closed set of commutation relations among those generators.

In Sec. 5.2, it is shown that the group $SL(2,c)$, the group of two-by-two unimodular matrices shares the same algebra as the Lorentz group. It is thus easier to study the Lorentz group using these two-by-two matrices. In Sec. 5.3, the four-component space–time and momentum-energy four-vectors can be written in the form of 2×2 matrices. In Sec. 5.4, we study the subgroups of $SL(2,c)$, thus those of the Lorentz group. In Sec. 5.5, we study the transformation properties of $SL(2,c)$ matrices and of those for four-vectors.

The $SL(2,c)$ matrices have six independent parameters. Its subgroup, consisting only of real numbers, has three independent parameters. In Sec. 5.6, it is noted that this real matrix can be written as one boost matrix sandwiched between two rotation matrices. It is shown then that this becomes a triangular matrix which cannot be diagonalized.

5.1 Lie Algebra of the Lorentz Group

Let us start with the three dimensional space with x, y, and z coordinates. We can rotate this coordinate system around the z axis by writing

$$
\begin{pmatrix} \cos\phi & -\sin\phi & 0 \\ \sin\phi & \cos\phi & 0 \\ 0 & 0 & 1 \end{pmatrix} \begin{pmatrix} x \\ y \\ z \end{pmatrix} = \begin{pmatrix} (\cos\phi)x - (\sin\phi)y \\ (\sin\phi)x + (\cos\phi)y \\ z \end{pmatrix}. \tag{5.1}
$$

Thus, x and y become $(\cos\phi)x - (\sin\phi)y$ and $(\sin\phi)x + (\cos\phi)y$ respectively, while z remains unchanged. The matrix in this expression can be written as

$$
\exp\left[-i\phi L_3\right] = \sum_n \frac{(-i\phi L_3)^n}{n!} \tag{5.2}
$$

with

$$
L_3 = \begin{pmatrix} 0 & -i & 0 \\ i & 0 & 0 \\ 0 & 0 & 0 \end{pmatrix}. \tag{5.3}
$$

This matrix is called the generator of rotations around the z axis. Likewise, we can consider rotations around the x axis and also around the y axis. Their generators are

$$
L_1 = \begin{pmatrix} 0 & 0 & 0 \\ 0 & 0 & -i \\ 0 & i & 0 \end{pmatrix}, \quad \text{and} \quad L_2 = \begin{pmatrix} 0 & 0 & i \\ 0 & 0 & 0 \\ -i & 0 & 0 \end{pmatrix}, \tag{5.4}
$$

respectively.

These generators satisfy the commutation relations

$$
[L_1, L_2] = iL_3, \quad [L_2, L_3] = iL_1, \quad [L_3, L_1] = iL_2. \tag{5.5}
$$

These commutation relations can be written as

$$
[L_i, L_j] = i\epsilon_{ijk}L_k. \tag{5.6}
$$

This closed set of commutation relations is called the *Lie algebra* of the rotation group or of the three-dimensional rotation group.

There are other forms of operators that satisfy the same Lie algebra. The three operators

$$L_i = -i \left(x_j \frac{\partial}{\partial x_k} - x_k \frac{\partial}{\partial x_j} \right) \tag{5.7}$$

satisfy the Lie algebra given in Eq. (5.5) or Eq. (5.6). These operators are applicable to functions of the coordinate variables x, y, and z.

As for the Lorentz boost, let us go back to Eq. (2.6). The transformation given can be written as

$$\begin{pmatrix} 1 & 0 & 0 & 0 \\ 0 & 1 & 0 & 0 \\ 0 & 0 & \cosh\eta & \sinh\eta \\ 0 & 0 & \sinh\eta & \cosh\eta \end{pmatrix} \begin{pmatrix} x \\ y \\ z \\ t \end{pmatrix}. \tag{5.8}$$

We shall hereafter use the convention $c = 1$. The four-dimensional space of (x, y, z, t) is called the Minkowskian space. The generator for the Lorentz boost along the z direction is

$$K_3 = \begin{pmatrix} 0 & 0 & 0 & 0 \\ 0 & 0 & 0 & 0 \\ 0 & 0 & 0 & i \\ 0 & 0 & i & 0 \end{pmatrix}. \tag{5.9}$$

Likewise, we can consider boosts along the x and y directions, and their generators are

$$K_1 = \begin{pmatrix} 0 & 0 & 0 & i \\ 0 & 0 & 0 & 0 \\ 0 & 0 & 0 & 0 \\ i & 0 & 0 & 0 \end{pmatrix} \quad \text{and} \quad K_2 = \begin{pmatrix} 0 & 0 & 0 & 0 \\ 0 & 0 & 0 & i \\ 0 & 0 & 0 & 0 \\ 0 & i & 0 & 0 \end{pmatrix}, \tag{5.10}$$

respectively. The generators applicable to the function defined over the Minkowskian space are thus

$$K_i = -i \left(x_i \frac{\partial}{\partial t} + t \frac{\partial}{\partial x_i} \right). \tag{5.11}$$

In this Minkowskian space, the rotation matrix of Eq. (5.1) is expanded to the 4×4 matrix

$$\begin{pmatrix} \cos\phi & -\sin\phi & 0 & 0 \\ \sin\phi & \cos\phi & 0 & 0 \\ 0 & 0 & 1 & 0 \\ 0 & 0 & 0 & 1 \end{pmatrix} \begin{pmatrix} x \\ y \\ z \\ t \end{pmatrix} = \begin{pmatrix} (\cos\phi)x - (\sin\phi)y \\ (\sin\phi)x + (\cos\phi)y \\ z \\ t \end{pmatrix}, \tag{5.12}$$

and with its generator of Eq. (5.3) becomes the 4×4 matrix

$$J_3 = \begin{pmatrix} 0 & -i & 0 & 0 \\ i & 0 & 0 & 0 \\ 0 & 0 & 0 & 0 \\ 0 & 0 & 0 & 0 \end{pmatrix}. \qquad (5.13)$$

Likewise, the 3×3 matrices of L_1 and L_2 become expanded to

$$J_1 = \begin{pmatrix} 0 & 0 & 0 & 0 \\ 0 & 0 & -i & 0 \\ 0 & i & 0 & 0 \\ 0 & 0 & 0 & 0 \end{pmatrix} \quad \text{and} \quad J_2 = \begin{pmatrix} 0 & 0 & i & 0 \\ 0 & 0 & 0 & 0 \\ -i & 0 & 0 & 0 \\ 0 & 0 & 0 & 0 \end{pmatrix}, \qquad (5.14)$$

respectively. These 4×4 matrices satisfy the same Lie algebra given for the 3×3 matrices in Eq. (5.6):

$$[J_i, J_j] = i\epsilon_{ijk}J_k. \qquad (5.15)$$

Let us go back to the boost generators of Eqs. (5.9) and (5.10), and take commutation relations. Then

$$[K_i, K_j] = -i\epsilon_{ijk}J_k. \qquad (5.16)$$

Thus, those three boost generators cannot form a closed set of commutation relations. They need the rotation generators. If we take commutation relations between the J and K matrices,

$$[J_i, K_j] = i\epsilon_{ijk}K_k. \qquad (5.17)$$

The 4×4 J and K matrices are called the generators of the Lorentz group. They form the following closed set of commutation relations.

$$[J_i, J_j] = i\epsilon_{ijk}J_k, \quad [J_i, K_j] = i\epsilon_{ijk}K_k, \quad [K_i, K_j] = -i\epsilon_{ijk}J_k. \qquad (5.18)$$

This set of commutation relations is called the Lie algebra of the Lorentz group.

5.2 2×2 Representation of the Lorentz Group

Let us consider the 2×2 matrices

$$J_1 = \frac{1}{2}\sigma_1, \quad J_2 = \frac{1}{2}\sigma_2, \quad J_3 = \frac{1}{2}\sigma_3, \qquad (5.19)$$

where

$$\sigma_1 = \begin{pmatrix} 0 & 1 \\ 1 & 0 \end{pmatrix}, \quad \sigma_2 = \begin{pmatrix} 0 & -i \\ i & 0 \end{pmatrix}, \quad \sigma_3 = \begin{pmatrix} 1 & 0 \\ 0 & -1 \end{pmatrix}. \qquad (5.20)$$

These 2×2 matrices are called the Pauli spin matrices. They are Hermitian and their role in physics is well known.

In addition, let us consider the following three anti-Hermitian matrices.

$$K_1 = \frac{i}{2}\sigma_1, \quad K_2 = \frac{i}{2}\sigma_2, \quad K_3 = \frac{i}{2}\sigma_3. \tag{5.21}$$

Then, these matrices satisfy the Lie algebra of the Lorentz group given in Eq. (5.18). They generate a group of 2×2 unimodular matrices. The determinant of every unimodular matrix is one. This group is known as $SL(2, c)$ or the special linear group with complex elements. Possible physical applications of this group are extensively discussed in the literature (Dirac, 1945a; Bargmann, 1947; Naimark, 1954; Wigner, 1960a,b; Kim and Noz, 1986; Başkal *et al.*, 2014, 2015)

Let us write the most general form for this matrix as

$$\begin{pmatrix} \alpha & \beta \\ \gamma & \delta \end{pmatrix}, \tag{5.22}$$

where all the elements are complex numbers. There are thus eight real numbers. Since the determinant of this matrix is to be one, there are only six independent numbers. Another way of expressing this matrix with six independent numbers is to write it in the form

$$\exp\left\{ -i \sum_i (\theta_i J_i + \lambda_i K_i) \right\} \tag{5.23}$$

with the six real parameters of θ_i and λ_i. The Taylor expansion of this exponential form will result in a 2×2 matrix like that of Eq. (5.22).

This $SL(2, c)$ matrix is the smallest matrix containing as many as six degrees of freedom. For this reason, it occupies an important place in mathematics with applications in many branches of physics. For instance, 2×2 matrices play the central role in optical sciences (Başkal *et al.*, 2015).

Let us go back to the Lorentz group. It is remarkable that this 2×2 representation shares the same Lie algebra with the 4×4 representation for the Lorentz group. If two groups share the same Lie algebra, they are *locally isomorphic to each other* in the language of group theory. However, we shall use a more convenient language. We shall simply say one is *like* the other.

Indeed, the Lorentz group is like the group $SL(2, c)$. For this reason, it is said that $SL(2, c)$ is the covering group of the Lorentz group. In this book, we shall simply say that $SL(2, c)$ is the 2×2 representation of the Lorentz group.

For each 2×2 matrix, there is a corresponding 4×4 matrix, and it is possible to write the most general form of the 4×4 matrix in terms of the complex parameters α, β, γ and δ (Kim and Noz, 1986). However, for the purpose of this book, it is not necessary to give this complicated expression. We thus list only the special cases that are useful for the purpose of this book. They are given in Table 5.1.

Table 5.1   Generators and transformation matrices of $SL(2,c)$, and their corresponding 4×4 transformation matrices in the Lorentz group. The 4×4 matrices are applicable to the Minkowskian space of x, y, z, t.

Generators	2×2	4×4
$J_1 = \frac{1}{2}\begin{pmatrix} 0 & 1 \\ 1 & 0 \end{pmatrix}$	$\begin{pmatrix} \cos(\theta/2) & i\sin(\theta/2) \\ i\sin(\theta/2) & \cos(\theta/2) \end{pmatrix}$	$\begin{pmatrix} 1 & 0 & 0 & 0 \\ 0 & \cos\theta & -\sin\theta & 0 \\ 0 & \sin\theta & \cos\theta & 0 \\ 0 & 0 & 0 & 1 \end{pmatrix}$
$K_1 = \frac{1}{2}\begin{pmatrix} 0 & i \\ i & 0 \end{pmatrix}$	$\begin{pmatrix} \cosh(\lambda/2) & \sinh(\lambda/2) \\ \sinh(\lambda/2) & \cosh(\lambda/2) \end{pmatrix}$	$\begin{pmatrix} \cosh\lambda & 0 & 0 & \sinh\lambda \\ 0 & 1 & 0 & 0 \\ 0 & 0 & 1 & 0 \\ \sinh\lambda & 0 & 0 & \cosh\lambda \end{pmatrix}$
$J_2 = \frac{1}{2}\begin{pmatrix} 0 & -i \\ i & 0 \end{pmatrix}$	$\begin{pmatrix} \cos(\theta/2 & -\sin(\theta/2) \\ \sin(\theta/2) & \cos(\theta/2) \end{pmatrix}$	$\begin{pmatrix} \cos\theta & 0 & \sin\theta & 0 \\ 0 & 1 & 0 & 0 \\ -\sin\theta & 0 & \cos\theta & 0 \\ 0 & 0 & 0 & 1 \end{pmatrix}$
$K_2 = \frac{1}{2}\begin{pmatrix} 0 & 1 \\ -1 & 0 \end{pmatrix}$	$\begin{pmatrix} \cosh(\lambda/2) & -i\sinh(\lambda/2) \\ i\sinh(\lambda/2) & \cosh(\lambda/2) \end{pmatrix}$	$\begin{pmatrix} 1 & 0 & 0 & 0 \\ 0 & \cosh\lambda & 0 & \sinh\lambda \\ 0 & 0 & 1 & 0 \\ 0 & \sinh\lambda & 0 & \cosh\lambda \end{pmatrix}$
$J_3 = \frac{1}{2}\begin{pmatrix} 1 & 0 \\ 0 & -1 \end{pmatrix}$	$\begin{pmatrix} \exp(i\phi/2) & 0 \\ 0 & \exp(-i\phi/2) \end{pmatrix}$	$\begin{pmatrix} \cos\phi & -\sin\phi & 0 & 0 \\ \sin\phi & \cos\phi & 0 & 0 \\ 0 & 0 & 1 & 0 \\ 0 & 0 & 0 & 1 \end{pmatrix}$
$K_3 = \frac{1}{2}\begin{pmatrix} i & 0 \\ 0 & -i \end{pmatrix}$	$\begin{pmatrix} \exp(\eta/2) & 0 \\ 0 & \exp(-\eta/2) \end{pmatrix}$	$\begin{pmatrix} 1 & 0 & 0 & 0 \\ 0 & 1 & 0 & 0 \\ 0 & 0 & \cosh\eta & \sinh\eta \\ 0 & 0 & \sinh\eta & \cosh\eta \end{pmatrix}$

5.3 Four-vectors in the 2 × 2 Representation

In the three-dimensional Euclidean space, the coordinate variables x, y, and z define one vector. Likewise, in the four-dimensional Minkowskian space, the coordinate variables x, y, z, and t constitute one set of four numbers. It is thus appropriate to define the vector in this four-dimensional space and call it the *four-vector*. It is more convenient to re-order the coordinate variables as (t, x, y, z) and the energy–momentum four-vector (p_0, p_x, p_y, p_z), with $p_0 = E$.

In the 4×4 Minkowskian space of (t, x, y, z), we know how to apply 4×4 matrices to perform transformations. For Lorentz transformations, the quantity

$$t^2 - x^2 - y^2 - z^2 \qquad (5.24)$$

remains constant. Likewise, the energy–momentum four-vector

$$p_0^2 - p_x^2 - p_y^2 - p_z^2 \qquad (5.25)$$

is invariant under Lorentz transformations.

In physics, the dot product $p \cdot x$ is an important Lorentz-invariant quantity, where

$$p \cdot x = p_0 t - p_x x - p_y y - p_z z. \qquad (5.26)$$

This is a scalar product of two four-vectors.

The question is how to write these four-vectors in the 2×2 formalism. Yes, those 2×2 matrices that are applicable to two-component column vectors are called *spinors*. The question is how to construct the four-vectors from the two-component $SL(2, c)$ spinors. These questions were addressed in the literature on the Lorentz group (Kim and Noz, 1986).

We can skip those mathematical details and start with the Hermitian matrix

$$[X] = \begin{pmatrix} t + z & x - iy \\ x + iy & t - z \end{pmatrix}. \qquad (5.27)$$

The determinant of this matrix is the Lorentz-invariant quantity given in Eq. (5.24). The energy–momentum four-vector can thus be written as

$$[P] = \begin{pmatrix} p_0 + p_z & p_x - ip_y \\ p_x + ip_y & p_0 - p_z \end{pmatrix}. \qquad (5.28)$$

The determinant of this matrix leads to the invariant quantity of Eq. (5.25).

Let us write the matrix

$$[P + X] = \begin{pmatrix} p_0 + p_z + t + z & p_x - ip_y + x - iy \\ p_x + ip_y + x + iy & p_0 - p_z + t - z \end{pmatrix}. \tag{5.29}$$

The determinant of this expression gives (Wigner, 1960b; Başkal *et al.*, 2015)

$$p \cdot x = \frac{1}{2} \left(\det[P + X] - \det[P] - \det[X] \right). \tag{5.30}$$

Let us go to Table 5.1, and pick the 2×2 boost matrix along the z direction, and name it as $B(\eta)$:

$$B(\eta) = \begin{pmatrix} e^{\eta/2} & 0 \\ 0 & e^{-\eta/2} \end{pmatrix}. \tag{5.31}$$

If we take the matrix product $B(\eta)XB^\dagger(\eta)$, the result is

$$B(\eta)[X]B^\dagger(\eta) = \begin{pmatrix} (t+z)e^\eta & x - iy \\ x + iy & (t-z)e^{-\eta} \end{pmatrix} \tag{5.32}$$

resulting in the transformation

$$t \rightarrow (\cosh \eta)t + (\sinh \eta)z, \quad \text{and} \quad z \rightarrow (\sinh \eta)t + (\cosh \eta)z.$$

This is clearly a Lorentz boost along the z direction.

Again from Table 5.1, we pick the rotation matrix around the z axis and write it as

$$Z(\phi) = \begin{pmatrix} e^{-i\phi/2} & 0 \\ 0 & e^{i\phi/2} \end{pmatrix}. \tag{5.33}$$

Then

$$Z(\phi)[X]Z^\dagger(\phi) = \begin{pmatrix} t+z & (x - iy)e^{-i\phi} \\ (x + iy)e^{i\phi} & t - z \end{pmatrix} \tag{5.34}$$

resulting in the rotation around the z axis:

$$x \rightarrow (\cos \phi)x - (\sin \phi)y \quad \text{and} \quad y \rightarrow (\sin \phi)x + (\cos \phi)y.$$

Finally, let us consider a rotation around the axis perpendicular to the z axis, and pick the rotation around the y axis. The rotation matrix takes the form

$$R(\theta) = \begin{pmatrix} \cos(\theta/2) & -\sin(\theta/2) \\ \sin(\theta/2) & \cos(\theta/2) \end{pmatrix}. \tag{5.35}$$

If we apply this matrix to the four-vector X in the same manner as in the cases of $B(\eta)$ of $Z(\phi)$, the result is

$$R(\theta)[X]R^\dagger(\theta) = \begin{pmatrix} t + z' & x' - iy \\ x' + iy & t - z' \end{pmatrix}, \qquad (5.36)$$

with

$$\begin{pmatrix} z' \\ x' \end{pmatrix} = \begin{pmatrix} \cos\theta & -\sin\theta \\ \sin\theta & \cos\theta \end{pmatrix} \begin{pmatrix} z \\ x \end{pmatrix}.$$

The Lie algebra of the three-dimensional rotation group given in Eq. (5.5) tells us that the rotations around two orthogonal directions can lead to a rotation around the third axis. The Lie algebra of Eq. (5.18) tells us the Lorentz boost along one direction can be applied to all directions. Thus, the three transformations given in Eqs.(5.32), (5.34), and (5.36) define the most general transformations in the four-dimensional Minkowskian space.

5.4 Subgroups of the Lorentz Group

There are nine commutations relations in Eq. (5.18). These commutation relations are invariant under Hermitian conjugation. Among the six generators, the rotation generators J_i remain invariant under Hermitian conjugation, while the boost generators K_i are anti-Hermitian with $K_i^* = -K_i$. This means that, for every Lie algebra of Eq. (5.18), there is another Lie algebra where K_i is replaced by

$$\dot{K}_i = -K_i. \qquad (5.37)$$

Thus, to every Lie algebra of the Lorentz group, there corresponds its dotted algebra leading to the dotted representation. These two groups are therefore subgroups to each other.

In this section, we are interested in subsets of those nine commutation relations which can be grouped into a closed set. The rotation generators J_i satisfy the closed set of three commutation relations. Therefore, the rotation group is a subgroup of the Lorentz group. These generators are Hermitian while the boost generators are not. Thus, the rotation subgroup is the Hermitian subgroup of the Lorentz group.

Let us now restrict our attention to the 2×2 representation. Among the six generators, K_1, K_3, and J_2 are pure imaginary. They thus give matrices with real elements. Furthermore, they satisfy the following set of commutation relations.

$$[K_1, K_3] = iJ_2, \quad [J_2, K_3] = iK_1, \quad [J_2, K_1] = -iK_3. \qquad (5.38)$$

This subgroup is like the Lorentz group consisting of the rotation around the y and boosts along the x and z directions. This group serves many useful purposes in physics. It is known as the two-dimensional symplectic group or $Sp(2)$ in the literature (Guillemin and Sternberg, 2001).

We can also consider the following combinations of the generators.

$$J_3, \quad N_1 = K_1 - J_2, \quad N_2 = K_2 + J_1, \tag{5.39}$$

with their explicit expressions:

$$J_3 = \frac{1}{2}\begin{pmatrix} 1 & 0 \\ 0 & -1 \end{pmatrix}, \quad N_1 = \begin{pmatrix} 0 & i \\ 0 & 0 \end{pmatrix}, \quad N_2 = \begin{pmatrix} 0 & 1 \\ 0 & 0 \end{pmatrix}. \tag{5.40}$$

These generators satisfy the following closed set of commutation relations.

$$[N_1, N_2] = 0, \quad [J_3, N_1] = iN_2, \quad [J_3, N_2] = -iN_1. \tag{5.41}$$

Unlike the rotation subgroup and the subgroup generated by Eq. (5.38), it is difficult to construct a geometry corresponding to this Lie algebra.

However, Wigner considered two-dimensional Euclidean transformations consisting of one rotation and translations along the two orthogonal directions (Wigner, 1939). In the two-dimensional space of x and y, the rotation generator is

$$J_3 = -i\left(x\frac{\partial}{\partial y} - y\frac{\partial}{\partial x} \right), \tag{5.42}$$

and the translation generators are

$$P_1 = -i\frac{\partial}{\partial x} \quad \text{and} \quad P_2 = -i\frac{\partial}{\partial y}. \tag{5.43}$$

Then

$$[P_1, P_2] = 0, \quad [J_3, P_1] = iP_2, \quad [J_3, P_2] = -iP_1. \tag{5.44}$$

This set of commutation relations is the same as the set given in Eq. (5.41). Thus, the subgroup generated by the Lie algebra of Eq. (5.41) is like (locally isomorphic to) the two-dimensional Euclidean group (Wigner, 1939).

5.5 Transformation Properties in the 2 × 2 Representation

In the 4×4 representation, transformations are straightforward: apply 4×4 matrices to the four-component vector. On the other hand, in the 2×2 representation, the four–vector is written in terms of a two-by-two matrix. The momentum-energy four-vector is written in the form given in Eq. (5.28).

If the particle is at rest, the momentum-energy four–vector becomes

$$[P] = \begin{pmatrix} m & 0 \\ 0 & m \end{pmatrix}. \tag{5.45}$$

Let us rotate this four-vector around the z and y directions. Since these rotation matrices are Hermitian

$$Z(\phi)[P]Z^\dagger(\phi) = R(\theta)[P]R^\dagger(\theta) = P, \tag{5.46}$$

where the 2×2 matrices for $Z(\phi)$ and $R(\theta)$ are given in Eqs. (5.33) and (5.35), respectively. This expression can also be written as

$$Z(\phi)[P]Z^{-1}(\phi) = R(\theta)[P]R^{-1}(\theta) = [P], \tag{5.47}$$

because the rotation matrices are Hermitian.

On the other hand, for the boost matrix of Eq. (5.31), its Hermitian conjugate is not its inverse. It remains invariant. Thus,

$$B(\eta)[P]B^\dagger = \begin{pmatrix} e^{\eta/2} & 0 \\ 0 & e^{-\eta/2} \end{pmatrix} \begin{pmatrix} m & 0 \\ 0 & m \end{pmatrix} \begin{pmatrix} e^{\eta/2} & 0 \\ 0 & e^{-\eta/2} \end{pmatrix} = \begin{pmatrix} me^\eta & 0 \\ 0 & me^{-\eta} \end{pmatrix}, \tag{5.48}$$

while $B(\eta)[P]B^{-1} = [P]$ is

$$\begin{pmatrix} e^{\eta/2} & 0 \\ 0 & e^{-\eta/2} \end{pmatrix} \begin{pmatrix} m & 0 \\ 0 & m \end{pmatrix} \begin{pmatrix} e^{-\eta/2} & 0 \\ 0 & e^{\eta/2} \end{pmatrix} = \begin{pmatrix} m & 0 \\ 0 & m \end{pmatrix}. \tag{5.49}$$

This difference comes from the fact that not all the matrices of $SL(2,c)$ are Hermitian.

5.6 Decompositions of the $SL(2,c)$ Matrices

Among the subgroups of $SL(2,c)$, the rotation subgroup is used most often in physics. It is still largely unknown that the subgroup consisting of real matrices plays important roles in many branches of physics, particularly in particle physics and optical sciences (Başkal *et al.*, 2015). The matrix in this subgroup has three degrees of freedom, and thus can be decomposed into

$$\begin{pmatrix} \cos(\theta_1/2) & -\sin(\theta_1/2) \\ \sin(\theta_1/2) & \cos(\theta_1/2) \end{pmatrix} \begin{pmatrix} \cosh\lambda & \sinh\lambda \\ \sinh\lambda & \cosh\lambda \end{pmatrix} \begin{pmatrix} \cos(\theta_2/2) & -\sin(\theta_2/2) \\ \sin(\theta_2/2) & \cos(\theta_2/2) \end{pmatrix}. \tag{5.50}$$

In this expression, a Lorentz boost along the x direction is sandwiched between two matrices for rotation around the y direction. This decomposition is called the Bargmann decomposition (Bargmann, 1947).

This form leads to interesting and useful results when $\theta_1 = \theta_2 = \theta$. Then the matrix multiplication leads to

$$\begin{pmatrix} (\cos\theta)\cosh\lambda & \sinh\lambda - (\sin\theta)\cosh\lambda \\ \sinh\lambda + (\sin\theta)\cosh\lambda & (\cos\theta)\cosh\lambda \end{pmatrix}. \tag{5.51}$$

When $\tanh\lambda = \sin\theta$, this matrix becomes

$$\begin{pmatrix} 1 & 0 \\ 2\sinh\lambda & 1 \end{pmatrix}. \tag{5.52}$$

This form is called the Iwasawa decomposition (Iwasawa, 1949). This triangular matrix cannot be diagonalized, but can explain a very important aspect of physics as we shall see in Chapters 6 and 7.

Chapter 6

Wigner's Little Groups

In 1963, Eugene Paul Wigner was awarded the Nobel Prize in Physics. A photo of the Nobel ceremony is shown in Fig. 6.1. The prize was for his contributions to the theory of the atomic nucleus and the elementary particles, particularly through the discovery and application of fundamental symmetry principles.

There are no disputes about this statement. On the other hand, there still is a question of why the statement did not mention Wigner's 1939 paper on the Lorentz group (Wigner, 1939), which was regarded by Wigner and many others as his most important contribution in physics. The reason was very simple. To most of the physicists at that time, including his departmental colleagues at Princeton, this paper appeared to be a mathematical exposition having nothing to do with physics. Yet, there were a number of physicists who attempted to connect the paper with the physical world. Steven Weinberg was one of them, and published, after 1963, a series of papers in the Physical Review (Weinberg, 1964a,b,c). In Fig. 6.1, he is talking to Wigner at a meeting of Princeton graduate students in 1957.

One of the authors (YSK) of this book started studying this paper in 1960 while he was a graduate student at Princeton. While he continued his interest in this paper, he was advised by his colleagues not to waste time. After all, Wigner did not get the Nobel Prize for this paper, and thus it is worthless. This advice was given by some of the distinguished physicists who were associated with Wigner.

(Photo from Martha Wigner)   (Photo by Dieter Brill)

Fig. 6.1.   Eugene Paul Wigner and King of Sweden at the 1963 Nobel Ceremony, and Steven Weinberg talking to Wigner while he was a graduate student at Princeton in 1957.

Those distinguished physicists had some specific reasons. They noted that Wigner introduced the subgroups of the Lorentz group whose transformations do not change the momentum of a given particle — the momentum remains invariant. Those subgroups are known as Wigner's *little groups* (Wigner, 1939). First, they contended that Wigner's little groups cannot explain the Dirac equation for spin-1/2 particles. Second, the little groups cannot explain the Maxwell four-vector and four-tensor for electromagnetic waves.

As for the Dirac equation, they were too lazy to study the problem. As for the Maxwell case, the problem was not completely settled until 1990 when Wigner published a paper with one of the present authors (Kim and Wigner, 1990b), even though Weinberg in 1964 constructed gauge-invariant four-tensors for the electromagnetic field starting from Wigner's 1939 paper (Weinberg, 1964c).

We examine in this chapter the nature of the gaps that existed between Wigner's little groups and the physics world. After closing those gaps, we are able to position Wigner at the proper place in the map of physicists, as illustrated in Fig. 6.2.

In 1979, the physics world celebrated the Einstein centennial year. He was born in 1879. Many people wrote articles and produced art works. Among them was a portrait of Wigner and Einstein produced in 1978, as shown in Fig. 6.3. This portrait becomes meaningful to us after understanding the full content of Wigner's little groups which allow us to interpret the internal space–time symmetries of particles in Einstein's Lorentz-covariant world.

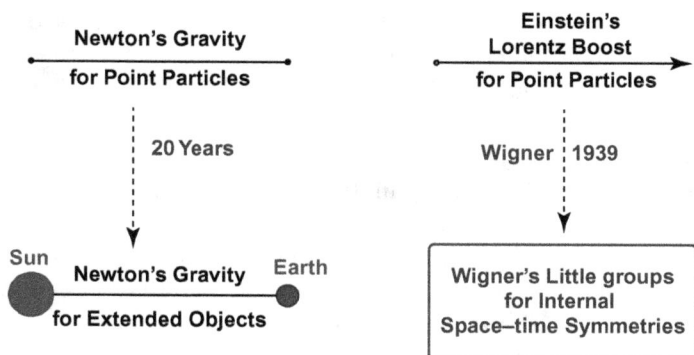

Fig. 6.2. Wigner's 1939 paper on internal space–time symmetries. Let us go back to two figures given in the Preface. We can combine those figures into one. As the Lorentz group provided the mathematical tool for Einstein's formulation of special relativity, Wigner's mathematics provides the tools for studying internal space–time symmetries of particles in the Lorentz-covariant world.

Fig. 6.3. Wigner with Einstein. This portrait was constructed by Bulent Atalay in 1978 for the Einstein Centennial Year of 1979.

6.1 Introduction

The purpose of the present chapter is to discuss whether Wigner's 1939 paper on the Lorentz group provides the framework to address the internal space–time symmetries of particles in the Lorentz-covariant world. This question is far more important than whether Wigner deserved a Nobel Prize for this paper alone.

For many years since 1963, many people claimed that Wigner's 1939 paper is worthless because he did not get the Nobel Prize for it. Let us respond to this fatalistic view. Einstein did not get the prize for his formulation of special relativity in 1905. Does this mean that Einstein's special relativity is worthless?

However, it is quite possible that Wigner started this subject, but did not finish it. If so, how did this happen? In his 1939 paper (Wigner, 1939), Wigner considered the subgroups of the Lorentz group whose transformations leave the momentum of a given particle invariant. These subgroups are called Wigner's *little groups*. It was not clear to the physics world that the little groups dictate the internal space–time symmetries in the Lorentz-covariant world.

Wigner observed first that a massive particle at rest has three rotational degrees of freedom leading to the concept of spin. Thus, the little group for this massive particle is like $O(3)$. How about this massive particle moving in the z direction? We could settle this issue easily.

Wigner observed also that a massless particle cannot be brought to its rest frame, but he showed that the little group for the massless particle also has three degrees of freedom, and that this little group is locally isomorphic to the group $E(2)$ or the two-dimensional Euclidean group. This means that generators of this little group share the same Lie algebra with the two-dimensional Euclidean group with one rotational and two translational degrees of freedom.

It is not difficult to associate the rotational degree of freedom of $E(2)$ to the helicity of the massless particle. However, what is the physics of those two translational degrees of freedom? Wigner did not provide the answer to this question in his original paper (Wigner, 1939). There is an opinion in the physics community that Wigner did not get his Nobel Prize for his 1939 paper because of this.

To be more specific, let us go to Fig. 6.4, showing the pages from the 1939 paper. There are two 4×4 matrices. One of them is for the Lorentz boost along the z direction. The other is for the little group for massless particles. Wigner did not offer a physical interpretation of this strange matrix.

Indeed, this question has a stormy history. In 1964, Weinberg constructed a representation space independent of this matrix, and ended up with the gauge-independent electromagnetic field tensor (Weinberg, 1964c). In 1971, (Janner and Janssen, 1971) constructed this matrix for a gauge

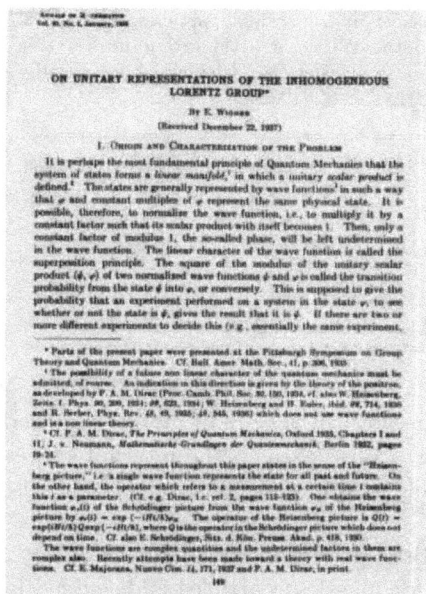

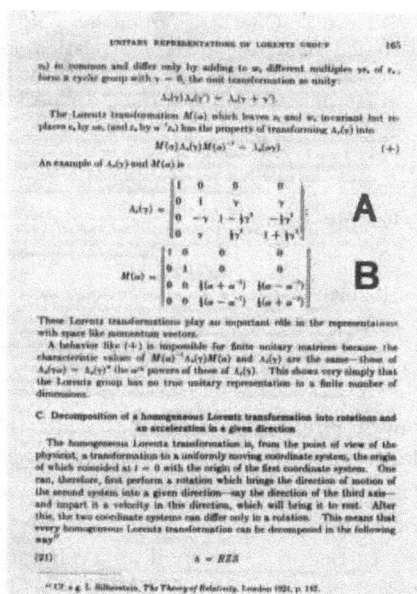

Fig. 6.4. Wigner's 1939 paper in the Annals of Mathematics. Its front page is on page 149 of the journal. On page 165, there are two strange matrices. The second matrix (matrix B) is for the Lorentz boost along the z direction. However, the physics of the first matrix (matrix A) was not completely understood until 1990, 51 years after 1939. (Courtesy of the Annals of Mathematics.)

transformation of electromagnetic waves interacting with electrons. In 1976, Kupersztych obtained the same expression by considering a rotation followed by a Lorentz boost which leaves the photon momentum invariant and concluded that this matrix performs a gauge transformation when applied to the electromagnetic four-vector, without mentioning Wigner's little group for massless particles (Kupersztych, 1976). In 1981, Han and Kim considered a rotation and two boosts whose resulting transformation would leave the photon four-momentum invariant. They also noted that this matrix performs a gauge transformation. In addition, they pointed out the matrix was constructed by Wigner for his little group for photons (Han and Kim, 1981). However, the issue was not completely settled until 1990 (Kim and Wigner, 1990b), 51 years after 1939, or 27 years after his Nobel Prize in 1963.

In this chapter, we point out that the complete understanding of this matrix leads to the result given in Table 6.1. As Einstein's

Table 6.1   One little group for both massive and massless particles. Einstein's special relativity gives one relation for both. Wigner's little group unifies the internal space–time symmetries for massive and massless particles which are locally isomorphic to $O(3)$ and $E(2)$, respectively (Han *et al.*, 1983).

	Massive slow	Lorentz covariance	Fast massless
Energy–momentum	$mc^2 + p^2/2m$	Einstein's $$E = \sqrt{(mc^2)^2 + (cp)^2}$$	$E = cp$
Helicity spin, gauge	S_3 S_1, S_2	Wigner's little group	Helicity gauge trans.

energy–momentum leads to expressions both in the small-momentum and large-momentum limits, Wigner's little groups explain the internal space–time symmetries for the massive particle at rest as well as for the massless particle.

In Sec. 6.2, we spell out Wigner's little groups in the language of 2×2 matrices. In Sec. 6.3, the little group for massless particles is discussed in detail. In Sec. 6.4, the 2×2 representation is given for spin-1/2 particles. The gauge transformation is defined for this 2×2 representation. In Sec. 6.5, the $O(3)$-like little group for massive particles is presented. In Sec. 6.6, we discuss the continuity problem for massive, massless, and imaginary mass particles.

The Dirac equation is for the internal space–time symmetries of spin-1/2 particles in the Lorentz-covariant world. We shall study the Dirac equation and Dirac matrices in detail in Chapter 7.

6.2   Wigner's Little Groups

In his 1939 paper (Wigner, 1939), Wigner noted that the most important quantities for a given particle are its mass and momentum, thus, its four-momentum. In addition to the four-momentum, what other variables are needed for the complete description of the particle?

If the particle is at rest, its momentum is zero, but it has spin with three different directions. A light wave has electric and magnetic fields perpendicular to the direction of propagation. This is translated into the photon spin parallel or anti-parallel to the direction of motion. Massless spin-half neutrinos are polarized. These are the issues concerning the internal space–time symmetries.

Thus, Wigner considered the subgroups of the Lorentz group whose transformations do not change the four-momentum of a given particle. These subgroups are called Wigner's little groups.

Let us choose z axis as the direction of the momentum. Then, in the 2×2 representation, the four-momentum takes the form

$$[P] = \begin{pmatrix} E+p & 0 \\ 0 & E-p \end{pmatrix}, \tag{6.1}$$

where p is the magnitude of momentum. The determinant of this matrix is $E^2 - p^2$, which is m^2. It is a Lorentz-invariant quantity.

If this particle is at rest, the momentum four-vector becomes

$$m \begin{pmatrix} 1 & 0 \\ 0 & 1 \end{pmatrix}. \tag{6.2}$$

If the particle is massless, the four-momentum is

$$2p \begin{pmatrix} 1 & 0 \\ 0 & 0 \end{pmatrix}. \tag{6.3}$$

We do not observe particles moving faster than light, but they play roles in physical theories. Their masses are imaginary with negative values of m^2. Thus, $p^2 = -m^2$ and the momentum four-vector can be written as

$$p \begin{pmatrix} 1 & 0 \\ 0 & -1 \end{pmatrix}. \tag{6.4}$$

Since the momentum has three degrees of freedom, and since the momentum is fixed, each little group has three degrees of freedom. First of all, all three of these matrices are invariant under rotations around the z axis, since

$$Z(\phi)[P]Z^{\dagger}(\phi) = [P] \tag{6.5}$$

where $[P]$ is given in Eqs. (6.2)–(6.4). The rotation matrix $Z(\phi)$ takes the form

$$Z(\phi) = \begin{pmatrix} e^{-i\phi/2} & 0 \\ 0 & e^{i\phi/2} \end{pmatrix}, \tag{6.6}$$

as given in Table 5.1. This rotation matrix is generated by

$$J_3 = \frac{1}{2} \begin{pmatrix} 1 & 0 \\ 0 & -1 \end{pmatrix}. \tag{6.7}$$

Thus, in addition to this degree of freedom, there are two additional degrees of freedom. For the massive particle at rest, the momentum four-vector is proportional to the identity matrix. Therefore, the rotation matrix $R(\theta)$ satisfies the condition

$$R(\theta)[P]R^\dagger(\theta) = [P] \tag{6.8}$$

where $[P]$ is given in Eq. (6.2), and $R(\theta)$ takes the form

$$\begin{pmatrix} \cos(\theta/2) & -\sin(\theta/2) \\ \sin(\theta/2) & \cos(\theta/2) \end{pmatrix}. \tag{6.9}$$

Since the rotation axis can be rotated around the z axis, rotations around the y axis leave the four-momentum invariant. Indeed, the little group for the massive particle at rest are like $O(3)$ or the three-dimensional rotation group.

For a massless particle moving along the z direction, the little group should satisfy

$$D[P]D^\dagger = [P] \tag{6.10}$$

where $[P]$ is given in Eq. (6.3). The 2×2 D matrix should be of the form

$$D(\gamma, \xi) = \begin{pmatrix} 1 & \gamma - i\xi \\ 0 & 1 \end{pmatrix} \tag{6.11}$$

with two real parameters γ and ξ. This D matrix can be written as

$$\exp\left(-i\gamma N_1 - i\xi N_2\right) \tag{6.12}$$

with

$$N_1 = K_1 - J_2 = \frac{1}{2}\begin{pmatrix} 0 & i \\ 0 & 0 \end{pmatrix},$$

$$N_2 = K_2 + J_1 = \frac{1}{2}\begin{pmatrix} 0 & 1 \\ 0 & 0 \end{pmatrix}. \tag{6.13}$$

It was shown in Sec. 5.4 that these two matrices, together with the rotation generator of Eq. (6.7), satisfy the set of commutation relations which are like the two-dimensional Euclidean group. This was observed first by Wigner in 1939 (Wigner, 1939). We shall discuss the physics of N_1 and N_2 in Sec. 6.3.

As for the little group whose transformations leave the four-momentum of Eq. (6.4) invariant, we can consider the matrix

$$S(\lambda) = \begin{pmatrix} \cosh(\lambda/2) & \sinh(\lambda/2) \\ \sinh(\lambda/2) & \cosh(\lambda/2) \end{pmatrix}. \tag{6.14}$$

This matrix corresponds to the Lorentz boost along the x direction. This matrix is generated by K_1 listed in Table 5.1. Then this matrix satisfies the condition

$$S(\lambda)[P]S^{\dagger}(\lambda) = [P], \tag{6.15}$$

where $[P]$ is the four-momentum matrix of Eq. (6.4). Since there is a rotational degree around the z axis, the Lorentz-boost along the y direction will lead to the same result.

6.3 Massless Particles

Let us go back to N_1 and N_2, introduced in Sec. 6.2 as well as Sec. 5.4. The $D(\gamma, \xi)$ matrix is constructed not to change the momentum–energy four-vector of the massless particle moving along the z direction. Then we are led to examine what happens when this matrix is applied to the electromagnetic four-vector, which can be written as

$$[A] = \begin{pmatrix} A_0 + A_z & A_x - iA_y \\ A_x + iA_y & A_0 - A_z \end{pmatrix}. \tag{6.16}$$

This matrix can be transformed in the way the space–time and momentum four-vectors are transformed in the Lorentzian space. However, unlike the momentum–energy four-vector of Eq. (6.3), this matrix has off-diagonal elements. The transformation in this case can be written as

$$D(\gamma, \xi)[A]D^{\dagger}(\gamma, \xi), \tag{6.17}$$

and the matrix multiplication leads to

$$[A] + 2 \begin{pmatrix} \gamma A_x + \xi A_y & 0 \\ 0 & 0 \end{pmatrix} + (A_0 - A_z) \begin{pmatrix} \gamma^2 + \xi^2 & \gamma - i\xi \\ \gamma + i\xi & 0 \end{pmatrix}. \tag{6.18}$$

If we place the Lorentz condition $A_0 = A_z$, then $[A]$ becomes

$$[A] = \begin{pmatrix} A_0 + A_z & A_x - iA_y \\ A_x + iA_y & 0 \end{pmatrix} \tag{6.19}$$

and the D transformed $[A]$ of Eq. (6.18) becomes

$$[A] + 2 \begin{pmatrix} \gamma A_x + \xi A_y & 0 \\ 0 & 0 \end{pmatrix}. \tag{6.20}$$

This means that A_x and A_y are not changed, but the quantity

$$(\gamma A_x + \xi A_y)$$

is added to A_0 and to A_z. In other words, the D matrix performs a gauge transformation.

Yes, the 2×2 matrix D was generated by the Lie algebra for the two-dimensional Euclidean group with two independent translational degrees of freedom. On the other hand, the D transformation given so far changes only one parameter. In order to examine this result further, let us study the geometry of the Euclidean transformation in the three dimensional space of (x, y, z). The rotation around the z axis is generated by

$$J_3 = \begin{pmatrix} 0 & -i & 0 \\ i & 0 & 0 \\ 0 & 0 & 0 \end{pmatrix}, \quad P_1 = \begin{pmatrix} 0 & 0 & i \\ 0 & 0 & 0 \\ 0 & 0 & 0 \end{pmatrix}, \quad P_2 = \begin{pmatrix} 0 & 0 & 0 \\ 0 & 0 & i \\ 0 & 0 & 0 \end{pmatrix}. \tag{6.21}$$

These matrices of course satisfy the Lie algebra given in Eq. (5.44). The commutation relations are

$$[P_1, P_2] = 0, \quad [J_3, P_1] = iP_2, \quad [J_3, P_2] = -iP_1, \tag{6.22}$$

and the transformation matrix becomes

$$E(\gamma, \xi) = \exp\left(-i\gamma P_1 - i\xi P_2\right) = \begin{pmatrix} 1 & 0 & \gamma \\ 0 & 1 & \xi \\ 0 & 0 & 1 \end{pmatrix}. \tag{6.23}$$

When this matrix is applied to the (x, y, z) coordinates with $z = 1$, the result is translations along the x and y variables:

$$\begin{pmatrix} 1 & 0 & \gamma \\ 0 & 1 & \xi \\ 0 & 0 & 1 \end{pmatrix} \begin{pmatrix} x \\ y \\ 1 \end{pmatrix} = \begin{pmatrix} x + \gamma \\ y + \xi \\ 1 \end{pmatrix}. \tag{6.24}$$

The $E(\gamma, \xi)$ indeed performs translations on the x, y plane.

If we take the Hermitian conjugate of the E matrix of Eq. (6.23), the result is

$$C(\gamma, \xi) = \begin{pmatrix} 1 & 0 & 0 \\ 0 & 1 & 0 \\ \gamma & \xi & 1 \end{pmatrix}. \tag{6.25}$$

This matrix can be written as

$$C(\gamma, \xi) = \exp\left(-i\gamma Q_1 - i\xi Q_2\right) = \begin{pmatrix} 1 & 0 & 0 \\ 0 & 1 & 0 \\ \gamma & \xi & 1 \end{pmatrix}, \tag{6.26}$$

with

$$Q_1 = \begin{pmatrix} 0 & 0 & 0 \\ 0 & 0 & 0 \\ i & 0 & 0 \end{pmatrix}, \qquad Q_2 = \begin{pmatrix} 0 & 0 & 0 \\ 0 & 0 & 0 \\ 0 & i & 0 \end{pmatrix}. \tag{6.27}$$

These matrices satisfy the same Lie algebra as that for P_1 and P_2:

$$[Q_1, Q_2] = 0, \quad [J_3, Q_1] = iQ_2, \quad [J_3, Q_2] = -iQ_1. \tag{6.28}$$

On the other hand, in the differential form, they are quite different from the translation generators. They can be written as

$$Q_1 = -ix\frac{\partial}{\partial z} \quad \text{and} \quad Q_2 = -iy\frac{\partial}{\partial z}. \tag{6.29}$$

These operators do not change the x and y coordinates. They add to the z coordinate quantities proportional to values of x and y. As is shown in Fig. 6.5, this is a transformation on a cylindrical surface. Thus, we call the group generated by the Lie algebra of Eq. (6.28) the *cylindrical group*. The geometry of internal space–time symmetry of massless particles is that of a cylinder with one rotational degree of freedom corresponding to the helicity and one translational degree of freedom along the side surface of the cylinder (Kim and Wigner, 1987). This geometry is shown in Fig. 6.5.

6.4 Spin-1/2 Particles

Let us go back to the Lie algebra of the Lorentz group given in Eq. (5.18). It was noted that there are six 4×4 matrices satisfying nine commutation

Euclidean   **Cylindrical**

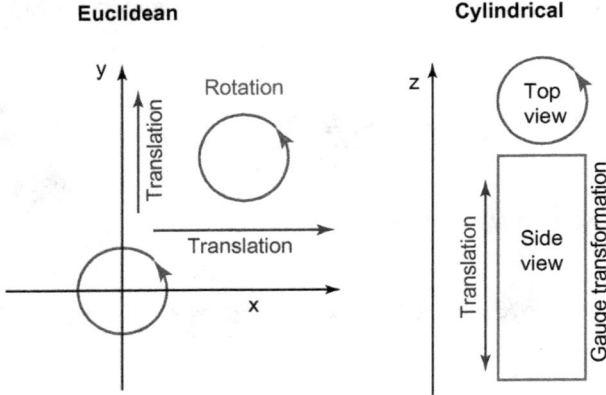

Fig. 6.5.   The Euclidean group (left) has one rotational degree of freedom and two translational degrees of freedom. The cylindrical group has one rotational degree of freedom viewed from the top and one translational degree of freedom viewed along the longitudinal surface. This translational degree of freedom can be associated with gauge transformations (Kim and Wigner, 1987).

relations. It is possible to construct the same Lie algebra with six 2×2 matrices (Kim and Noz, 1986). They are

$$J_i = \frac{1}{2}\sigma_i \quad \text{and} \quad K_i = \frac{i}{2}\sigma_i, \tag{6.30}$$

where σ_i are the Pauli spin matrices. While J_i are Hermitian, K_i are not. They are anti-Hermitian. Since the Lie algebra of Eq. (5.18) is Hermitian invariant, we can construct the same Lie algebra with

$$J_i = \frac{1}{2}\sigma_i \quad \text{and} \quad \dot{K}_i = -\frac{i}{2}\sigma_i. \tag{6.31}$$

This is the reason why the 4×4 Dirac matrices can explain both the spin-1/2 particle and the anti-particle.

Thus, the most general form of the transformation matrix takes the form

$$T = \exp\left(-\frac{i}{2}\sum_i \theta_i\sigma_i + \frac{1}{2}\sum_i \eta_i\sigma_i\right), \tag{6.32}$$

and this transformation matrix is applicable to the spinors

$$u = \begin{pmatrix} 1 \\ 0 \end{pmatrix} \quad \text{and} \quad v = \begin{pmatrix} 0 \\ 1 \end{pmatrix}. \tag{6.33}$$

In addition, we have to consider the transformation matrices

$$\dot{T} = \exp\left(-\frac{i}{2}\sum_i \theta_i \sigma_i - \frac{1}{2}\sum_i \eta_i \sigma_i\right), \tag{6.34}$$

applicable to

$$\dot{u} = \begin{pmatrix} 1 \\ 0 \end{pmatrix} \quad \text{and} \quad \dot{v} = \begin{pmatrix} 0 \\ 1 \end{pmatrix}. \tag{6.35}$$

With this understanding, let us go back to the Lie algebra of Eq. (5.18). Here again the rotation generators satisfy the closed set of commutation relations:

$$[J_i, J_j] = i\epsilon_{ijk}J_k, \qquad \left[\dot{J}_i, \dot{J}_j\right] = i\epsilon_{ijk}\dot{J}_k. \tag{6.36}$$

These operators generate the rotation-like $SU(2)$ group, whose physical interpretation is well known, namely the electron and positron spins.

Here also we can consider the $E(2)$-like subgroup generated by

$$J_3, \qquad N_1 = K_1 - J_2, \qquad N_2 = K_2 + J_1. \tag{6.37}$$

The N_1 and N_2 matrices take the form

$$N_1 = \begin{pmatrix} 0 & i \\ 0 & 0 \end{pmatrix}, \qquad N_2 = \begin{pmatrix} 0 & 1 \\ 0 & 0 \end{pmatrix}. \tag{6.38}$$

On the other hand, in the *dotted* representation,

$$\dot{N}_1 = \begin{pmatrix} 0 & 0 \\ -i & 0 \end{pmatrix}, \qquad \dot{N}_2 = \begin{pmatrix} 0 & 0 \\ 1 & 0 \end{pmatrix}. \tag{6.39}$$

There are therefore two different D matrices:

$$D(\gamma, \xi) = \exp\left\{-(i\gamma N_1 + i\xi N_2)\right\} = \begin{pmatrix} 1 & \gamma - i\xi \\ 0 & 1 \end{pmatrix}, \tag{6.40}$$

and

$$\dot{D}(\gamma, \xi) = \exp\left\{-\left(i\gamma\dot{N}_1 + i\xi\dot{N}_2\right)\right\} = \begin{pmatrix} 1 & 0 \\ \gamma + i\xi & 1 \end{pmatrix}. \tag{6.41}$$

These are the gauge transformation matrices applicable to massless spin-1/2 particles (Han *et al.*, 1982, 1986b).

The spinors u and $\dot{v}$ are gauge-invariant since

$$D(\gamma, \xi)u = u, \quad \text{and} \quad \dot{D}(\gamma, \xi)\dot{v} = \dot{v}. \tag{6.42}$$

As for v and $\dot{u}$,

$$D(\gamma, \xi)v = v + (\gamma - i\xi)u,$$

$$\dot{D}(\gamma, \xi)\dot{u} = \dot{u} + (\gamma + i\xi)\dot{v}. \tag{6.43}$$

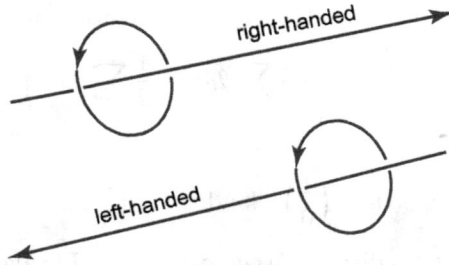

Fig. 6.6. Polarization of massless neutrinos. Massless neutrinos are left-handed, while anti-neutrinos are right-handed. This is a consequence of gauge invariance.

They are not invariant under the D transformations, and they are not gauge-invariant. Thus, we can conclude that the polarization of massless neutrinos is a consequence of gauge invariance, as illustrated in Fig. 6.6.

6.5 Massless Particle as a Limiting Case of a Massive Particle

In 1953, Inönü and Wigner started with a sphere with the symmetry of the group $O(3)$ or the three-dimensional rotation group (Inönü and Wigner, 1953). They observed that the surface of the sphere becomes flat when the radius of the sphere becomes large, and considered a flat surface tangent to the sphere.

Then the group theory of the rotation group will become that of the flat surface in the large-radius limit. Let us start with the generators of the rotation group:

$$J_x = -iy\frac{\partial}{\partial z} + iz\frac{\partial}{\partial y}, \quad J_y = -iz\frac{\partial}{\partial x} + ix\frac{\partial}{\partial z},$$
$$J_z = -ix\frac{\partial}{\partial y} + iy\frac{\partial}{\partial x}. \tag{6.44}$$

Near the north pole, x and y are much smaller than z, and z is almost constant. Within this range of variables,

$$\frac{1}{R}J_x = +i\frac{\partial}{\partial y} \quad \text{and} \quad \frac{1}{R}J_y = -i\frac{\partial}{\partial x}, \tag{6.45}$$

where R is the radius of the sphere, while J_z remains unchanged. These are the generators of the Euclidean group. Inönü and Wigner called this limiting process the *contraction* of $O(3)$ to $E(2)$. They used this contraction procedure for obtaining the Galilean group for slow particles from the Lorentz group.

It was not until 1983 when Han *et al.* observed that the $O(3)$-like little group for massive particles can be contracted to the $E(2)$-like little group for massless particles, while the radius of the sphere corresponds to the speed of the particle (Han *et al.*, 1983).

In 1987, Kim and Wigner considered the equatorial belt where z is much smaller than x and y (Kim and Wigner, 1987). Then J_x and J_y of Eq. (6.5) become

$$J_x = -iy\frac{\partial}{\partial z}, \qquad J_y = ix\frac{\partial}{\partial z}, \tag{6.46}$$

while J_z is not changed. These are the generators of the cylindrical group applicable to massless particles with one rotational degree of freedom and one gauge degree of freedom.

The purpose of this section is to Lorentz boost Wigner's $O(3)$-like little group and show that this boosted $O(3)$ becomes the $E(2)$-like little group in the infinite-momentum limit, as illustrated in Fig. 6.7.

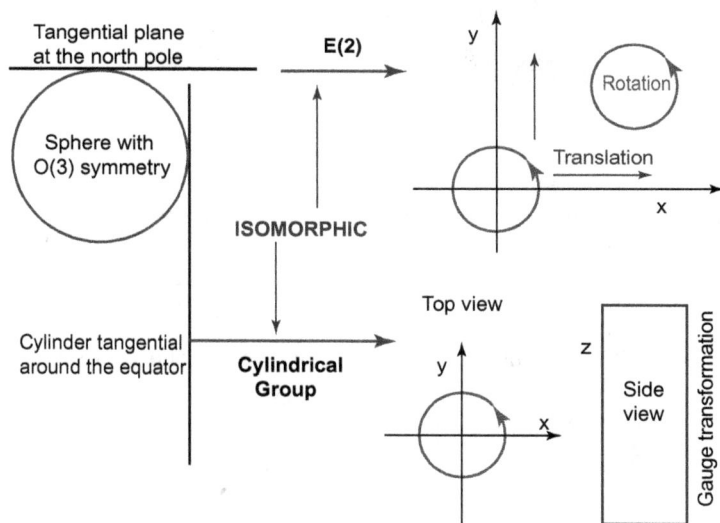

Fig. 6.7. $O(3)$-like and $E(2)$-like internal space–time symmetries of massive and massless particles. The sphere corresponds to the $O(3)$-like little group for the massive particle. There is a plane tangential to the sphere at its north pole which is $E(2)$. There is also a cylinder tangent to the sphere at its equatorial belt. This cylinder gives one helicity and one gauge degree of freedom. This figure thus gives a unified picture of the little groups for massive and massless particles (Başkal *et al.*, 2015).

In the 2×2 representation, the Lorentz boost along the positive direction is

$$B(\eta) = \begin{pmatrix} e^{\eta/2} & 0 \\ 0 & e^{-\eta/2} \end{pmatrix}, \tag{6.47}$$

and the rotation around the y axis is

$$R(\theta) = \begin{pmatrix} \cos(\theta/2) & -\sin(\theta/2) \\ \sin(\theta/2) & \cos(\theta/2) \end{pmatrix}. \tag{6.48}$$

Then, the boosted rotation matrix is

$$B(\eta)R(\theta)B(-\eta) = \begin{pmatrix} \cos(\theta/2) & -e^{\eta}\sin(\theta/2) \\ e^{-\eta}\sin(\theta/2) & \cos(\theta/2) \end{pmatrix}. \tag{6.49}$$

If η becomes very large, and this matrix is to remain finite, θ has to become very small, and this expression becomes (Han *et al.*, 1983; Başkal *et al.*, 2015)

$$\begin{pmatrix} 1 - \rho^2 e^{-2\eta}/2 & \rho \\ -\rho e^{-2\eta} & 1 - \rho^2 e^{-2\eta}/2 \end{pmatrix}, \tag{6.50}$$

with

$$\rho = -\frac{1}{2}\theta e^{\eta}, \tag{6.51}$$

and ρ is positive when θ is negative. This expression becomes

$$D(\rho) = \begin{pmatrix} 1 & \rho \\ 0 & 1 \end{pmatrix}. \tag{6.52}$$

In this 2×2 representation, the rotation around the z axis is

$$Z(\phi) = \begin{pmatrix} e^{-i\phi/2} & 0 \\ 0 & e^{i\phi/2} \end{pmatrix}. \tag{6.53}$$

Thus

$$D(\gamma, \xi) = Z(\phi)D(\rho, 0)Z^{-1}(\phi), \tag{6.54}$$

which becomes

$$D(\gamma, \xi) = \begin{pmatrix} 1 & \gamma - i\xi \\ 0 & 1 \end{pmatrix}, \tag{6.55}$$

with

$$\gamma = \rho \cos \phi, \quad \text{and} \quad \xi = \rho \sin \phi. \tag{6.56}$$

Here, we have studied how the $O(3)$-like little group for the massive particle becomes the $E(2)$-like little group for the massless particle in the infinite-η limit. What does this limit mean physically? The parameter η can be derived from the speed of of the particle. We know $\tanh(\eta) = v/c$, where v is the speed of the particle. Then

$$\tanh \eta = \frac{p}{\sqrt{m^2 + p^2}}, \tag{6.57}$$

where m and p are the mass and the momentum of the particle, respectively. If m is much smaller than p,

$$e^{\eta} = \frac{\sqrt{2}p}{m}, \tag{6.58}$$

which becomes large when m becomes very small. Thus, the limit of large η means the zero-mass limit.

Let us carry out the same limiting process for the 4×4 representation. From the generators of the Lorentz group, it is possible to construct the 4×4 matrices for rotations around the y axis and Lorentz boosts along the z axis as (Başkal *et al.*, 2015)

$$R(\theta) = \exp\left(-i\theta J_2\right) \quad \text{and} \quad B(\eta) = \exp\left(-i\eta K_3\right), \tag{6.59}$$

respectively. The Lorentz-boosted rotation matrix is $B(\eta)R(\theta)B(-\eta)$ which can be written as

$$\begin{pmatrix} \cos\theta & 0 & (\sin\theta)\cosh\eta & -(\sin\theta)\sinh\eta \\ 0 & 1 & 0 & 0 \\ -(\sin\theta)\cosh\eta & 0 & \cos\theta - b\sinh^2\eta & b(\cosh\eta)\sinh\eta \\ -(\sin\theta)\cosh\eta & 0 & -b(\cosh\eta)\sinh\eta & \cos\theta + b\cosh^2\eta \end{pmatrix}, \tag{6.60}$$

with $b = (1 - \cos\theta)$. While $\tanh\eta = v/c$, this boosted rotation matrix becomes a transformation matrix for a massless particle when η becomes infinite. On the other hand, if the matrix is to be finite in this limit, the angle θ has to become small. If we let $\rho = -\frac{1}{2}\theta e^{\eta}$ as given in Eq. (6.51), this 4×4 matrix becomes

$$
\begin{pmatrix}
1 & 0 & -\rho & \rho \\
0 & 1 & 0 & 0 \\
\rho & 0 & 1 - \rho^2/2 & \rho^2/2 \\
\rho & 0 & -\rho^2/2 & 1 + \rho^2/2
\end{pmatrix}.
\tag{6.61}
$$

This is the Lorentz-boosted rotation matrix around the y axis. However, we can rotate this y axis around the z axis by ϕ. Then the matrix becomes

$$
\begin{pmatrix}
1 & 0 & -\rho \cos \phi & \rho \cos \phi \\
0 & 1 & -\rho \sin \phi & \rho \sin \phi \\
\rho \cos \phi & \rho \sin \phi & 1 - \rho^2/2 & \rho^2/2 \\
\rho \cos \phi & \rho \sin \phi & -\rho^2/2 & 1 + \rho^2/2
\end{pmatrix}.
\tag{6.62}
$$

If we replace $\rho \cos \phi$ and $\rho \sin \phi$ with γ and ξ, respectively according to Eq. (6.56), this expression becomes

$$
\begin{pmatrix}
1 & 0 & -\gamma & \gamma \\
0 & 1 & -\xi & \xi \\
\gamma & \xi & 1 - \left(\gamma^2 + \xi^2\right)/2 & \left(\gamma^2 + \xi^2\right)/2 \\
\gamma & \xi & -\left(\gamma^2 + \xi^2\right)/2 & 1 + \left(\gamma^2 + \xi^2\right)/2
\end{pmatrix},
\tag{6.63}
$$

as shown in Fig. 6.4.

6.6   Continuity Problem

In this chapter, we studied different forms of matrices for the three little groups. The question is whether the little group for massive particles can be continued to that for massless particles, and then to that for imaginary-mass particles. For this purpose, we are led to consider the matrices

$$
\begin{pmatrix}
\cos(x) & -\sin(x) \\
\sin(x) & \cos(x)
\end{pmatrix}
\tag{6.64}
$$

for negative values of x, and

$$
\begin{pmatrix}
\cosh(x) & \sinh(x) \\
\sinh(x) & \cosh(x)
\end{pmatrix}
\tag{6.65}
$$

for positive values of x.

The question is how these matrices can be continued to each other. Another question is whether they become triangular during the continuation process. To answer these questions, let us Lorentz-boost them. They become

$$\begin{pmatrix} \cos(x) & -e^{\eta}\sin(x) \\ e^{-\eta}\sin(x) & \cos(x) \end{pmatrix} \quad \text{and} \quad \begin{pmatrix} \cosh(x) & e^{\eta}\sinh(x) \\ e^{-\eta}\sinh(x) & \cosh(x) \end{pmatrix}. \tag{6.66}$$

If the matrix is to remain finite for large values of η, $|x|$ has to be very small and the matrix should take the form

$$\begin{pmatrix} 1 - e^{-2\eta}\rho^2/2 & \rho \\ -e^{-2\eta}\rho & 1 - e^{-2\eta}\rho^2/2 \end{pmatrix}, \tag{6.67}$$

for small negative x, and

$$\begin{pmatrix} 1 + e^{-2\eta}\rho^2/2 & \rho \\ e^{-2\eta}\rho & 1 + e^{-2\eta}\rho^2/2 \end{pmatrix}, \tag{6.68}$$

for positive x, while $\rho = e^{\eta}|x|$ remains finite. In the extreme limit of large η, both matrices become triangular:

$$\begin{pmatrix} 1 & \rho \\ 0 & 1 \end{pmatrix}. \tag{6.69}$$

When a quantity is extremely close to zero, it can change its sign. This is how the matrix of Eq. (6.67) can make its continuation to that of Eq. (6.68), as illustrated in Fig. 6.8.

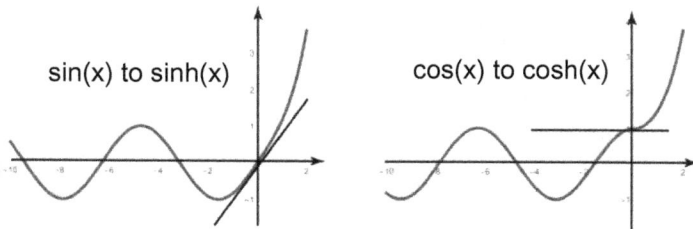

Fig. 6.8.   Continuities of functions. The functions $\sin(x)$ and $\cos(x)$ can be converted to $\sinh(x)$ and $\cosh(x)$, respectively at $x = 0$. Their first derivatives are continuous but the second derivatives are not. Thus, these are not analytic continuations. They are tangentially continuous.

Chapter 7

Lorentz Completion of the Little Groups

In his original paper (Wigner, 1939), Wigner worked out the three little groups with different Lie algebras and different sets of transformation matrices, depending on whether m^2 is positive, negative, or zero. Even though one little group can be continued to two different little groups as shown in Sec. 6.6, the continuity is not analytic. In this chapter, we shall show whether it is possible to study the same problem with a different choice of variables which will allow us to continue analytically to all the ranges of the mass variables.

While Wigner was interested in transformations that leave the momentum of a given particle invariant, he was not necessarily interested in fixing the mass variables. He was only interested in fixing the momentum variables. Let us thus consider a particle moving along the z direction with the magnitude of momentum p. Rotations round the z axis do not change this momentum.

This momentum can be rotated around the y axis. This rotated momentum is in the zx plane. It can then be Lorentz-boosted back to the original rotation. Thus, the net transformation is a rotation followed by a boost. Since this net transformation leaves the initial momentum invariant, it must be an element of Wigner's little group.

Rotations can be described by a circular arc, while Lorentz boosts will be straight lines. These line–circle combination should form a closed loop for the transformation that leaves the momentum invariant. Each circle and line should carry an arrow head to indicate the direction of transformation. Since the net transformation leaves the momentum invariant, it should be describable by a closed loop consisting of arcs and lines. We shall use this loop method to study Wigner's little groups in this chapter.

7.1   Introduction

In his original paper, Wigner worked out his little groups for the particle in a specific Lorentz frame. If the particle is massive, Wigner worked out his little group in the frame where the particle momentum is zero. When the particle mass is imaginary, he constructed the transformation matrix in the frame where $m^2 = -p^2$. Wigner did not discuss what happens when the momentum takes different values.

For a massive moving particle, it is possible to Lorentz-boost both the momentum and the little group matrix. It is even possible to boost the system to the infinite value of momentum to get the little group for massless particles, as indicated in Table 6.1.

In this chapter, we construct the transformation matrix whose parameters will cover all possible momentum and mass variables analytically by constructing loop transformations as described in Fig. 7.1. Since these loops preserve the original direction and magnitude of the momentum, it is appropriate to call them *Wigner loops*.

It was Kupersztych who showed in 1976 that it is possible to construct a momentum-preserving transformation by a rotation followed by a boost as

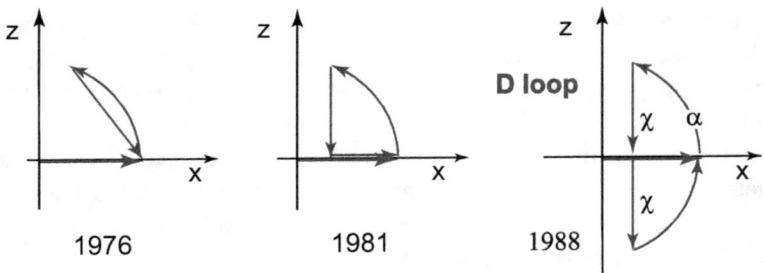

Fig. 7.1.   Evolution of the Wigner loop. In 1976, Kupersztych considered a rotation followed by a boost whose net result will leave the momentum invariant (Kupersztych, 1976). In 1981, Han and Kim considered the same problem with simpler forms for boost matrices (Han and Kim, 1981). In 1988, Han and Kim constructed the Lorentz kinematics (Han and Kim, 1988) corresponding to the Bargmann decomposition (Bargmann, 1947) consisting of one boost matrix sandwiched by two rotation matrices. In the present case, the two rotation matrices are identical.

shown in Fig. 7.1 (Kupersztych, 1976). He constructed his loop for a massless particle. In 1981, without knowing Kupersztych's earlier work, Han and Kim obtained the same result by constructing the loop with one rotation and two boosts as shown in Fig. 7.1 (Han and Kim, 1981). In 1988, Han and Kim showed that the same purpose can be achieved by one boost preceded and followed by the same rotation matrix, as shown also in Fig. 7.1 (Han and Kim, 1988). They showed that this process can be extended to particles with all possible values of mass. It is quite appropriate to call this loop *D loop*.

The D loop contains two circular arcs and one vertical line. They all carry their arrow heads. Thus, we can use the D loops to study the parity, time reversal, and charge conjugations of the little groups. This method is convenient for studying the symmetries of the Dirac equation.

In Sec. 7.2, we shall write down the 2×2 matrix for the D loop depending on two variables. It is shown that these variables cover analytically all possible masses. In Sec. 7.3, we discuss parity, time reversal, and charge conjugation in terms of the D loops. In Sec. 7.4, we discuss the Dirac matrices as a representation of the little group. In Sec. 7.5, we show the polarization of massless neutrinos, and in Sec. 7.6 we derive scalars, vectors, and tensors associated with the little groups.

7.2 Loop Representation of Wigner's Little Groups

In this section, we construct the 2×2 D matrix and study its properties. Let us assume that the momentum is along the z direction, the rotation around the z axis leaves the momentum invariant. According to the Euler decomposition (Han *et al.*, 1986a), the rotation around the y axis, in addition, will accommodate rotations along all three directions. For this reason, it is enough to study what happens in transformations within the xz plane (Han *et al.*, 1986a).

According to Fig. 7.1, the 2×2 matrix for the D loop can be written as

$$D(\alpha, \chi) = R(\alpha)S(-2\chi)R(\alpha). \tag{7.1}$$

The D matrix is written in terms of three transformations. This form is known in the literature as the Bargmann decomposition (Bargmann, 1947). This particular form gives an additional convenience. When we take the inverse or the Hermitian conjugate, we have to reverse the order of matrices. However, this form does not require re-ordering in the present case.

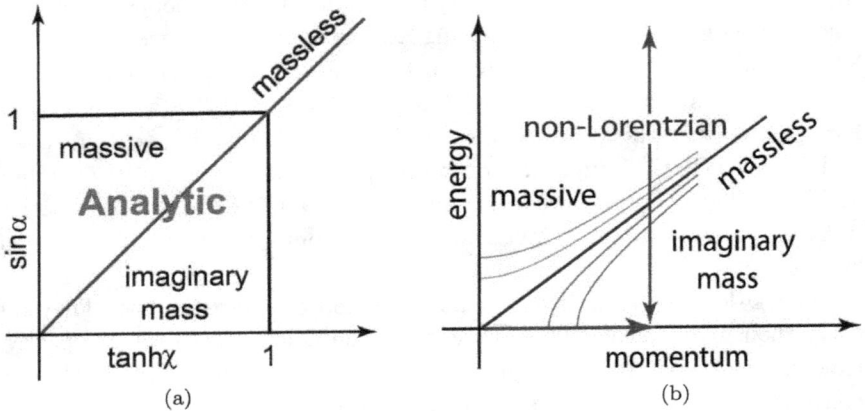

Fig. 7.2.   Non-Lorentzian transformations allowing mass variations. (a) The D matrix of Eq. (7.2) allows us to change the χ and α analytically within the square region. These variations allow the mass variations illustrated in (b), not allowed in Lorentz transformations. The Lorentz transformations are possible along the hyperbolas given in this figure.

After the matrix multiplication, this D matrix becomes

$$D(\alpha, \chi) = \begin{pmatrix} (\cos \alpha) \cosh \chi & -\sinh \chi - (\sin \alpha) \cosh \chi \\ -\sinh \chi + (\sin \alpha) \cosh \chi & (\cos \alpha) \cosh \chi \end{pmatrix}. \quad (7.2)$$

This matrix is analytic in both α and χ variables for all possible values as illustrated in Fig. 7.2(a). In this figure, the square region is divided into two triangular regions: one with $\sin \alpha > \tanh \chi$ and the other with $\sin \alpha < \tanh \chi$. The division is made by the line where $\sin \alpha = \tanh \chi$.

The question is whether it is possible to determine the mass of the particle from the parameters α, χ, and p. In Chapter 6, we noted that Wigner in his original paper worked out his little groups for three different regions of m^2, namely for $m^2 > 0, m^2 < 0$, and for $m^2 = 0$. For $m^2 = 0$, the little group matrix becomes triangular. Thus, the straight line $\sin \alpha = \tanh \chi$ in Fig. 7.2 corresponds to this region.

As for the massive particle with positive values of m, Wigner considered a particle at rest with four-momentum

$$\begin{pmatrix} m & 0 \\ 0 & m \end{pmatrix}. \quad (7.3)$$

He found out the Hermitian subgroup, namely the rotation subgroup, leaves this momentum invariant. Since the rotation around z belongs to this subgroup, it is then sufficient to consider the rotation around the y direction. The rotation matrix is

$$\begin{pmatrix} \cos(\theta/2) & -\sin(\theta/2) \\ \sin(\theta/2) & \cos(\theta/2) \end{pmatrix},$$ (7.4)

as was noted in Sec. 6.2. If we Lorentz-boost this matrix along the z direction, it becomes

$$\begin{pmatrix} \cos(\theta/2) & -e^{\eta}\sin(\theta/2) \\ e^{-\eta}\sin(\theta/2) & \cos(\theta/2) \end{pmatrix},$$ (7.5)

while the momentum matrix is boosted to

$$\begin{pmatrix} me^{\eta} & 0 \\ 0 & e^{-\eta}m \end{pmatrix} = \begin{pmatrix} \sqrt{m^2+p^2}+p & 0 \\ 0 & \sqrt{m^2+p^2}-p \end{pmatrix},$$ (7.6)

with

$$e^{-2\eta} = \frac{\sqrt{m^2+p^2}-p}{\sqrt{m^2+p^2}+p}.$$ (7.7)

Let us compare the transformation matrix of Eq. (7.5) with the D matrix of Eq. (7.2), and take the ratio

$$e^{-2\eta} = \frac{\text{lower-left element}}{\text{upper-right element}}.$$ (7.8)

Then

$$\frac{\sqrt{m^2+p^2}-p}{\sqrt{m^2+p^2}+p} = \frac{\sin\alpha - \tanh\chi}{\sin\alpha + \tanh\chi}.$$ (7.9)

This equation allows us to calculate m^2 in terms of α, χ, and p, and

$$m^2 = p^2 \left(\frac{\sin\alpha}{\tanh\chi} - 1 \right),$$ (7.10)

where m^2 is positive with $\sin\alpha > \tanh\chi$. The α and χ variables are within the upper triangular region in of Fig. 7.2(a).

If m^2 is negative, Wigner worked out his little group in the Lorentz frame where the momentum four-vector takes the form

$$\begin{pmatrix} m & 0 \\ 0 & -m \end{pmatrix},$$ (7.11)

and the little-group matrix is

$$\begin{pmatrix} \cosh(\lambda/2) & -\sinh(\lambda/2) \\ -\sinh(\lambda/2) & \cosh(\lambda/2) \end{pmatrix}. \tag{7.12}$$

We can Lorentz-boost both the four-momentum of Eq. (7.11) and the transformation matrix of Eq. (7.12), and compare them with the D matrix of Eq. (7.2) to calculate m^2, which becomes negative with $\sin\alpha < \tanh\chi$. The α and χ variables are now within the lower triangular region in Fig. 7.2(a).

The formula of Eq. (7.10) becomes zero when $\sin\alpha = \tanh\chi$. Thus, the same mass formula is valid for all possible values of $(\text{mass})^2$, and determined by the α and χ variables for a given value of p, as illustrated in Fig. 7.2(b).

7.3 Parity, Time Reversal, and Charge Conjugation

Space inversion leads to the sign change in χ:

$$D(\alpha, -\chi) = \begin{pmatrix} (\cos\alpha)\cosh\chi & \sinh\chi - (\sin\alpha)\cosh\chi \\ \sinh\chi + (\sin\alpha)\cosh\chi & (\cos\alpha)\cosh\chi \end{pmatrix}, \tag{7.13}$$

and time reversal leads to the sign change in both α and χ:

$$D(-\alpha, -\chi) = \begin{pmatrix} (\cos\alpha)\cosh\chi & \sinh\chi + (\sin\alpha)\cosh\chi \\ \sinh\chi - (\sin\alpha)\cosh\chi & (\cos\alpha)\cosh\chi \end{pmatrix}. \tag{7.14}$$

If we space-invert this expression, the result is a change only in the direction of rotation,

$$D(-\alpha, \chi) = \begin{pmatrix} (\cos\alpha)\cosh\chi & -\sinh\chi + (\sin\alpha)\cosh\chi \\ -\sinh\chi - (\sin\alpha)\cosh\chi & (\cos\alpha)\cosh\chi \end{pmatrix}. \tag{7.15}$$

The combined transformation of space inversion and time reversal is known as the *charge conjugation*. All these transformations are illustrated in Fig. 7.3.

Let us go back to the Lie algebra of Eq. (5.18) of Chapter 5. This algebra is invariant under Hermitian conjugation. This means that there is another set of commutation relations,

$$[J_i, J_j] = i\epsilon_{ijk}J_k, \qquad \left[J_i, \dot{K}_j\right] = i\epsilon_{ijk}\dot{K}_k, \qquad \left[\dot{K}_i, \dot{K}_j\right] = -i\epsilon_{ijk}J_k, \tag{7.16}$$

where K_i is replaced with $\dot{K}_i = -K_i$. These K_i generators are anti-Hermitian. According to the expressions given in Sec. 5.1, this transition

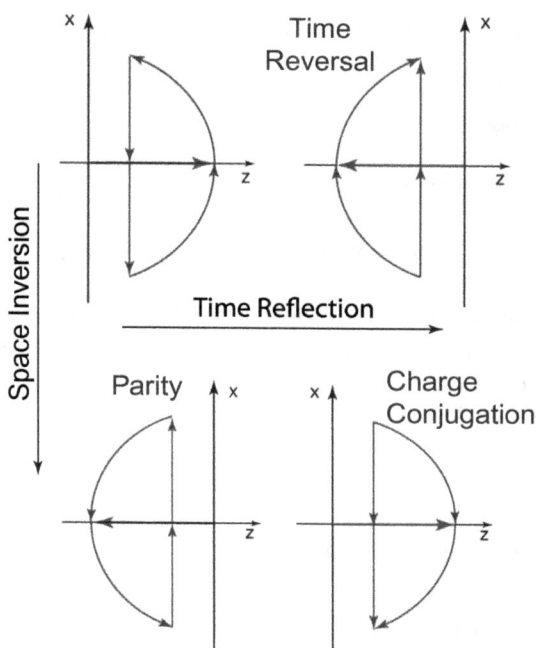

Fig. 7.3.   Parity, time reversal, and charge conjugation of Wigner's little groups in the loop representation.

to the dotted representation is achieved by the space inversion or by the parity operation.

On the other hand, the complex conjugation of the Lie algebra of Eq. (7.16) leads to

$$\left[J_i^*, J_j^*\right] = -i\epsilon_{ijk}J_k^*, \quad \left[J_i^*, K_j^*\right] = -i\epsilon_{ijk}K_k^*, \quad \left[K_i^*, K_j^*\right] = i\epsilon_{ijk}J_k^*.$$
(7.17)

It is possible to restore this algebra to that of the original form of Eq. (7.16) if we replace J_i^* by $-J_i$ and K_i^* by $-K_i$. This corresponds to the time-reversal process. This operation is known as the anti-unitary transformation in the literature (Wigner, 1960a,b).

Since the algebras of Eqs. (7.16) and (7.17) are invariant under the sign change of K_i and K_i^* respectively, there is another Lie algebra with J_i^* replaced by $-J_i$ and K_i^* by $-\dot{K}_i$. This is the parity operation followed by time reversal, resulting in charge conjugation. With the 4×4 matrices for spin-1 particles, this complex conjugation is trivial, and $J_i^* = -J_i$ as well as $K_i^* = -K_i$.

On the other hand for spin-1/2 particles, we note that

$$J_1^* = J_1, \qquad J_2^* = -J_2, \qquad J_3^* = J_3,$$

$$K_1^* = -K_1, \qquad K_2^* = K_2, \qquad K_3^* = -K_3. \qquad (7.18)$$

Thus, J_i^* should be replaced by $\sigma_2 J_i \sigma_2$, and K_i^* by $-\sigma_2 K_i \sigma_2$.

7.4 Dirac Matrices as a Representation of the Little Group

The Dirac equation, Dirac matrices, and Dirac spinors constitute the basic language for spin-1/2 particles in physics. Yet, they are not widely recognized as the package for Wigner's little group. Yes, the little group is for spins, so are the Dirac matrices.

Let us write the Dirac equation as

$$(p \cdot \gamma - m)\psi(\vec{x}, t) = \chi\psi(\vec{x}, t). \qquad (7.19)$$

This equation can be explicitly written as

$$\left(-i\gamma_0 \frac{\partial}{\partial t} - i\gamma_1 \frac{\partial}{\partial x} - i\gamma_2 \frac{\partial}{\partial y} - i\gamma_3 \frac{\partial}{\partial z} - m \right) \psi(\vec{x}, t) = \chi\psi(\vec{x}, t), \qquad (7.20)$$

where

$$\gamma_0 = \begin{pmatrix} 0 & I \\ I & 0 \end{pmatrix}, \quad \gamma_1 = \begin{pmatrix} 0 & \sigma_1 \\ -\sigma_1 & 0 \end{pmatrix}, \quad \gamma_2 = \begin{pmatrix} 0 & \sigma_2 \\ -\sigma_2 & 0 \end{pmatrix}, \quad \gamma_3 = \begin{pmatrix} 0 & \sigma_3 \\ -\sigma_3 & 0 \end{pmatrix}, \qquad (7.21)$$

where I is the 2×2 unit matrix. We use here the Weyl representation of the Dirac matrices.

The Dirac spinor has four components. Thus, we write the wave function for a free particle as

$$\psi(\vec{x}, t) = U_\pm \exp\left[i\left(\vec{p} \cdot \vec{x} - p_0 t\right)\right], \qquad (7.22)$$

with the Dirac spinors

$$U_+ = \begin{pmatrix} u \\ \dot{u} \end{pmatrix} \quad \text{and} \quad U_- = \begin{pmatrix} v \\ \dot{v} \end{pmatrix}, \qquad (7.23)$$

where

$$u = \dot{u} = \begin{pmatrix} 1 \\ 0 \end{pmatrix} \quad \text{and} \quad v = \dot{v} = \begin{pmatrix} 0 \\ 1 \end{pmatrix}. \qquad (7.24)$$

In Eq. (7.22), the exponential form $\exp\left[i\left(\vec{p} \cdot \vec{x} - p_0 t\right)\right]$ defines the particle momentum, and the column vector $U_\pm$ is for the representation space for

Wigner's little group dictating the internal space–time symmetries of spin-1/2 particles.

In this 4×4 representation, the generators for rotations and boosts take the form

$$J_i = \frac{1}{2} \begin{pmatrix} \sigma_i & 0 \\ 0 & \sigma_i \end{pmatrix}, \quad \text{and} \quad K_i = \frac{i}{2} \begin{pmatrix} \sigma_i & 0 \\ 0 & -\sigma_i \end{pmatrix}. \tag{7.25}$$

This means that both dotted and undotted spinors are transformed in the same way under rotation, while they are boosted in the opposite directions.

When this γ_0 matrix is applied to $U_\pm$

$$\gamma_0 U_+ = \begin{pmatrix} 0 & I \\ I & 0 \end{pmatrix} \begin{pmatrix} u \\ \dot{u} \end{pmatrix} = \begin{pmatrix} \dot{u} \\ u \end{pmatrix},$$

$$\gamma_0 U_- = \begin{pmatrix} 0 & I \\ I & 0 \end{pmatrix} \begin{pmatrix} v \\ \dot{v} \end{pmatrix} = \begin{pmatrix} \dot{v} \\ v \end{pmatrix}. \tag{7.26}$$

Thus, the γ_0 matrix interchanges the dotted and undotted spinors.

The 4×4 matrix for the rotation around the y axis is

$$R_{44}(\alpha) = \begin{pmatrix} R(\alpha) & 0 \\ 0 & R(\alpha) \end{pmatrix}, \tag{7.27}$$

while the matrix for the boost along the z direction is

$$B_{44}(\eta) = \begin{pmatrix} B(\eta) & 0 \\ 0 & B(-\eta) \end{pmatrix}, \tag{7.28}$$

with

$$B(\pm \eta) = \begin{pmatrix} e^{\pm \eta/2} & 0 \\ 0 & e^{\mp \eta/2} \end{pmatrix}. \tag{7.29}$$

These γ matrices satisfy the commutation relations

$$[\gamma_\mu, \gamma_\nu] = 2g_{\mu\nu}, \tag{7.30}$$

where

$$g_{00} = 1, \quad g_{11} = g_{22} = g_{22} = -1,$$

$$g_{\mu\nu} = 0 \quad \text{if} \quad \mu \neq \nu. \tag{7.31}$$

Let us consider space inversion with the exponential form changing to $\exp[i(-\vec{p} \cdot \vec{x} - p_0 t)]$. For this purpose, we can change the sign of x in the

Table 7.1 Parity, charge conjugation, and time reversal in the loop representation.

	Start	Time reflection
Start	Start with $R(\alpha)S(-2\chi)R(\alpha)$	Time reversal $R(-\alpha)S(2\chi)R(-\alpha)$
Space inversion	Parity $R(\alpha)S(2\chi)R(\alpha)$	Charge conjugation $R(-\alpha)S(-2\chi)R(-\alpha)$

Dirac equation of Eq. (7.20). It then becomes

$$\left(-i\gamma_0\frac{\partial}{\partial t} + i\gamma_1\frac{\partial}{\partial x} + i\gamma_2\frac{\partial}{\partial y} + i\gamma_3\frac{\partial}{\partial z} - m\right)\psi(-\vec{x},t) = \chi\psi(-\vec{x},t).$$
(7.32)

Since $\gamma_0\gamma_i = -\gamma_i\gamma_0$ for $i = 1, 2, 3$,

$$\left(-i\gamma_0\frac{\partial}{\partial t} - i\gamma_1\frac{\partial}{\partial x} - i\gamma_2\frac{\partial}{\partial y} - i\gamma_3\frac{\partial}{\partial z} - m\right)[\gamma_0\psi(-\vec{x}\cdot\vec{p}, p_0t)]$$
$$= \chi[\gamma_0\psi(-\vec{x}\cdot\vec{p}, p_0t)].$$
(7.33)

This is the Dirac equation under the space inversion or the parity operation. The Dirac spinor $U_\pm$ becomes $\gamma_0 U_\pm$, according to Eq. (7.26). This operation is illustrated in Table 7.1 and Fig. 7.3

We are interested in changing the sign of t. First, we can change both space and time variables, and then we can change the space variable. We can take the complex conjugate of the equation first. Since γ_2 is imaginary, while all others are real, the Dirac equation becomes

$$\left(i\gamma_0\frac{\partial}{\partial t} + i\gamma_1\frac{\partial}{\partial x} - i\gamma_2\frac{\partial}{\partial y} + i\gamma_3\frac{\partial}{\partial z} - m\right)\psi^*(\vec{x},t) = \chi\psi^*(\vec{x},t).$$
(7.34)

We are now interested in restoring this equation to the original form of Eq. (7.20). In order to achieve this goal, let us consider $(\gamma_1\gamma_3)$. This form commutes with γ_0 and γ_2, and anti-commutes with γ_1 and γ_3. Thus,

$$\left(i\gamma_0\frac{\partial}{\partial t} - i\gamma_1\frac{\partial}{\partial x} - i\gamma_2\frac{\partial}{\partial y} - i\gamma_3\frac{\partial}{\partial z} - m\right)(\gamma_1\gamma_3)\psi^*(\vec{x},t)$$
$$= \chi(\gamma_1\gamma_3)\psi^*(\vec{x},-t).$$
(7.35)

Furthermore, since

$$\gamma_1\gamma_3 = \begin{pmatrix} i\sigma_2 & 0 \\ 0 & i\sigma_2 \end{pmatrix}$$
(7.36)

this 4×4 matrix changes the direction of the spin. Indeed, this form of time reversal is consistent with Table 7.1 and Fig. 7.3.

Finally, let us change the signs of both $\vec{x}$ and t. For this purpose, we go back to the complex-conjugated Dirac equation of Eq. (7.34). Here, γ_2 anti-commutes with all others. Thus, the wave function

$$\gamma_2 \psi(-\vec{x} \cdot \vec{p}, -p_0 t) \tag{7.37}$$

should satisfy the Dirac equation. This form is known as the charge-conjugated wave function, and it is also illustrated in Table 7.1 and Fig. 7.3.

7.5 Polarization of Massless Neutrinos

For massless neutrinos, the little group consists of rotations around the z axis, in addition to N_i and $\dot{N}_i$ applicable to the upper and lower components of the Dirac spinors. Thus, the 4×4 matrix for these generators is

$$N_{44(i)} = \begin{pmatrix} N_i & 0 \\ 0 & \dot{N}_i \end{pmatrix}. \tag{7.38}$$

The transformation matrix is thus

$$D_{44}(\gamma, \xi) = \exp\left(-i\gamma N_{44(1)} - i\xi N_{44(2)}\right) = \begin{pmatrix} D(\gamma, \xi) & 0 \\ 0 & \dot{D}(\gamma, \xi) \end{pmatrix}, \tag{7.39}$$

with

$$D(\gamma, \xi) = \begin{pmatrix} 1 & \gamma - i\xi \\ 0 & 1 \end{pmatrix}, \qquad \dot{D}(\gamma, \xi) \begin{pmatrix} 1 & 0 \\ -\gamma - i\xi & 1 \end{pmatrix}. \tag{7.40}$$

As is illustrated in Fig. 6.7, the D transformation performs the gauge transformation on massless photons. Thus, this transformation allows us to extend the concept of gauge transformations to massless spin-1/2 particles. With this point in mind, let us see what happens when this D transformation is applied to the Dirac spinors.

$$D(\gamma, \xi)u = u, \qquad \dot{D}(\gamma, \xi)\dot{v} = \dot{v}. \tag{7.41}$$

Thus, u and $\dot{v}$ are invariant gauge transformations.

What happens to v and $\dot{u}$?

$$D(\gamma, \xi)v = v + (\gamma - i\xi)u, \qquad \dot{D}(\gamma, \xi)\dot{u} = \dot{u} - (\gamma + i\xi)\dot{v}. \tag{7.42}$$

These spinors are not invariant under gauge transformations (Han *et al.*, 1982, 1986b).

Thus, the Dirac spinor

$$U_{\text{inv}} = \begin{pmatrix} u \\ \dot{v} \end{pmatrix} \tag{7.43}$$

is gauge-invariant while the spinor

$$U_{\text{non}} = \begin{pmatrix} v \\ \dot{u} \end{pmatrix} \tag{7.44}$$

is not. Thus, gauge invariance leads to the polarization of massless spin-1/2 particles. Indeed, this is what we observe in the real world.

7.6 Scalars, Vectors, and Tensors

We are quite familiar with the process of constructing three spin-1 states and one spin-0 state from two spinors. Since each spinor has two states, there are four states if combined.

In the Lorentz-covariant world, for each spin-1/2 particle, there are two additional two-component spinors coming from the dotted representation (Kim and Noz, 1986; Başkal *et al.*, 2015; Weinberg, 1964a). There are thus four states. If two spinors are combined, there are 16 states. In this section, we show that they can be partitioned into

1. scalar with one state,
2. pseudo-scalar with one state,
3. four-vector with four states,
4. axial vector with four states,
5. second-rank tensor with six states.

These quantities contain 16 states. In our earlier publication (Başkal *et al.*, 2015), we did not take into account the parity operation properly. Here, we have completed the job (Başkal *et al.*, 2017).

For particles at rest, it is known that the addition of two one-half spins results in spin-0 and spin-1 states. Hence, we have two different spinors behaving differently under the Lorentz boost. Around the z direction, both spinors are transformed by

$$Z(\phi) = \exp\left(-i\phi J_3\right) = \begin{pmatrix} e^{-i\phi/2} & 0 \\ 0 & e^{i\phi/2} \end{pmatrix}. \tag{7.45}$$

However, they are boosted by

$$B(\eta) = \exp\left(-i\eta K_3\right) = \begin{pmatrix} e^{\eta/2} & 0 \\ 0 & e^{-\eta/2} \end{pmatrix},$$

$$\dot{B}(\eta) = \exp\left(i\eta K_3\right) = \begin{pmatrix} e^{-\eta/2} & 0 \\ 0 & e^{\eta/2} \end{pmatrix}, \tag{7.46}$$

which are applicable to the undotted and dotted spinors, respectively. These two matrices commute with each other, and also with the rotation matrix $Z(\phi)$ of Eq. (7.45). Since K_3 and J_3 commute with each other, we can work with the matrix $Q(\eta, \phi)$ defined as

$$Q(\eta, \phi) = B(\eta)Z(\phi) = \begin{pmatrix} e^{(\eta - i\phi)/2} & 0 \\ 0 & e^{-(\eta - i\phi)/2} \end{pmatrix},$$

$$\dot{Q}(\eta, \phi) = \dot{B}(\eta)\dot{Z}(\phi) = \begin{pmatrix} e^{-(\eta + i\phi)/2} & 0 \\ 0 & e^{(\eta + i\phi)/2} \end{pmatrix}. \tag{7.47}$$

When this combined matrix is applied to the spinors,

$$Q(\eta, \phi)u = e^{(\eta - i\phi)/2}u, \qquad Q(\eta, \phi)v = e^{-(\eta - i\phi)/2}v,$$

$$\dot{Q}(\eta, \phi)\dot{u} = e^{-(\eta + i\phi)/2}\dot{u}, \qquad \dot{Q}(\eta, \phi)\dot{v} = e^{(\eta + i\phi)/2}\dot{v}. \tag{7.48}$$

If the particle is at rest, we can explicitly construct the combinations

$$uu, \qquad \frac{1}{\sqrt{2}}(uv + vu), \qquad vv, \tag{7.49}$$

to obtain the spin-1 state, and

$$\frac{1}{\sqrt{2}}(uv - vu), \tag{7.50}$$

for the spin-0 state. This results in four bilinear states. In the $SL(2, c)$ regime, there are two dotted spinors which result in four more bilinear states. If we include both dotted and undotted spinors, there are 16 independent bilinear combinations. They are given in Table 7.2. This table also gives the effect of the operation of $Q(\eta, \phi)$.

Among the bilinear combinations given in Table 7.2, the following two equations are invariant under rotations and also under boosts:

$$S = \frac{1}{\sqrt{2}}(uv - vu) \quad \text{and} \quad \dot{S} = -\frac{1}{\sqrt{2}}(\dot{u}\dot{v} - \dot{v}\dot{u}). \tag{7.51}$$

They are thus scalars in the Lorentz-covariant world. Are they the same or different? Let us consider the following combinations:

$$S_+ = \frac{1}{\sqrt{2}}\left(S + \dot{S}\right) \quad \text{and} \quad S_- = \frac{1}{\sqrt{2}}\left(S - \dot{S}\right). \tag{7.52}$$

Under the dot conjugation, S_+ remains invariant, but S_- changes sign. The boost is performed in the opposite direction and therefore is the operation of space inversion. Thus, S_+ is a scalar while S_- is called a pseudo-scalar.

Table 7.2   Sixteen combinations of the $SL(2,c)$ spinors.

Spin 1	Spin 0
$uu, \quad \frac{1}{\sqrt{2}}(uv + vu), \quad vv,$	$\frac{1}{\sqrt{2}}(uv - vu)$
$\dot{u}\dot{u}, \quad \frac{1}{\sqrt{2}}(\dot{u}\dot{v} + \dot{v}\dot{u}), \quad \dot{v}\dot{v},$	$\frac{1}{\sqrt{2}}(\dot{u}\dot{v} - \dot{v}\dot{u})$
$u\dot{u}, \quad \frac{1}{\sqrt{2}}(u\dot{v} + v\dot{u}), \quad v\dot{v},$	$\frac{1}{\sqrt{2}}(u\dot{v} - v\dot{u})$
$\dot{u}u, \quad \frac{1}{\sqrt{2}}(\dot{u}v + \dot{v}u), \quad \dot{v}v,$	$\frac{1}{\sqrt{2}}(\dot{u}v - \dot{v}u)$

After the operation of $Q(\eta,\phi)$ and $\dot{Q}(\eta,\phi)$

$e^{-i\phi}e^{\eta}uu, \quad \frac{1}{\sqrt{2}}(uv + vu), \quad e^{i\phi}e^{-\eta}vv,$	$\frac{1}{\sqrt{2}}(uv - vu)$
$e^{-i\phi}e^{-\eta}\dot{u}\dot{u}, \quad \frac{1}{\sqrt{2}}(\dot{u}\dot{v} + \dot{v}\dot{u}), \quad e^{i\phi}e^{\eta}\dot{v}\dot{v},$	$\frac{1}{\sqrt{2}}(\dot{u}\dot{v} - \dot{v}\dot{u})$
$e^{-i\phi}u\dot{u}, \quad \frac{1}{\sqrt{2}}(e^{\eta}u\dot{v} + e^{-\eta}v\dot{u}), \quad e^{i\phi}v\dot{v},$	$\frac{1}{\sqrt{2}}(e^{\eta}u\dot{v} - e^{-\eta}v\dot{u})$
$e^{-i\phi}\dot{u}u, \quad \frac{1}{\sqrt{2}}(\dot{u}v + \dot{v}u), \quad e^{i\phi}\dot{v}v,$	$\frac{1}{\sqrt{2}}(e^{-\eta}\dot{u}v - e^{\eta}\dot{v}u)$

Notes: In the $SU(2)$ regime, there are two spinors leading to four bilinear forms. In the $SL(2,c)$ world, there are two undotted and two dotted spinors. These four spinors lead to 16 independent bilinear combinations.

7.6.1   *Four-vectors*

Let us go back to Eq. (7.49), and make a dot conjugation on one of the spinors.

$$u\dot{u}, \qquad \frac{1}{\sqrt{2}}(u\dot{v} + v\dot{u}), \qquad v\dot{v}, \qquad \frac{1}{\sqrt{2}}(u\dot{v} - v\dot{u}),$$

$$\dot{u}u, \qquad \frac{1}{\sqrt{2}}(\dot{u}v + \dot{v}u), \qquad \dot{v}v, \qquad \frac{1}{\sqrt{2}}(\dot{u}v - \dot{v}u). \qquad (7.53)$$

We can make symmetric combinations under dot conjugation which lead to:

$$\frac{1}{\sqrt{2}}(u\dot{u} + \dot{u}u), \quad \frac{1}{2}[(u\dot{v} + v\dot{u}) + (\dot{u}v + \dot{v}u)], \quad \frac{1}{\sqrt{2}}(v\dot{v} + \dot{v}v), \quad \text{for spin 1,}$$

$$\frac{1}{2}[(u\dot{v} - v\dot{u}) + (\dot{u}v - \dot{v}u)], \quad \text{for spin 0,} \qquad (7.54)$$

and anti-symmetric combinations which lead to:

$$\frac{1}{\sqrt{2}}\left(u\dot{u}-\dot{u}u\right),\quad \frac{1}{2}[(u\dot{v}+v\dot{u})-(\dot{u}v+\dot{v}u)],\quad \frac{1}{\sqrt{2}}(v\dot{v}-\dot{v}v),\quad \text{for spin 1,}$$

$$\frac{1}{2}[(u\dot{v}-v\dot{u})-(\dot{u}v-\dot{v}u)],\quad \text{for spin 0.} \tag{7.55}$$

Let us rewrite the expression for the space–time four-vector given in Eq. (5.27) as

$$\begin{pmatrix} t+z & x-iy \\ x+iy & t-z \end{pmatrix}, \tag{7.56}$$

which, under the parity operation, becomes

$$\begin{pmatrix} t-z & -x+iy \\ -x-iy & t+z \end{pmatrix}. \tag{7.57}$$

If the expression of Eq. (7.56) is for an axial vector, the parity operation leads to

$$\begin{pmatrix} -t+z & x-iy \\ x+iy & -t-z \end{pmatrix}, \tag{7.58}$$

where only the sign of t is changed. The off-diagonal elements remain invariant, while the diagonal elements are interchanged with sign changes.

We note here that the parity operation corresponds to dot conjugation. Then from the expressions given in Eqs. (7.54) and (7.55), it is possible to construct the four-vector as

$$V = \begin{pmatrix} u\dot{v}-\dot{v}u & v\dot{v}-\dot{v}v \\ u\dot{u}-\dot{u}u & \dot{u}v-v\dot{u} \end{pmatrix}, \tag{7.59}$$

where the off-diagonal elements change their signs under the dot conjugation, while the diagonal elements are interchanged.

The axial vector can be written as

$$A = \begin{pmatrix} u\dot{v}+\dot{v}u & v\dot{v}+\dot{v}v \\ u\dot{u}+\dot{u}u & -\dot{u}v-v\dot{u} \end{pmatrix}. \tag{7.60}$$

Here, the off-diagonal elements do not change their signs under dot conjugation and the diagonal elements become interchanged with a sign change. This matrix thus represents an axial vector.

7.6.2 *Second-rank Tensor*

There are also bilinear spinors which are both dotted or both undotted. We are interested in two sets of three quantities satisfying the $O(3)$ symmetry. They should therefore transform like

$$(x + iy)/\sqrt{2}, \qquad (x - iy)/\sqrt{2}, \qquad z, \tag{7.61}$$

which are like

$$uu, \qquad vv, \qquad (uv + vu)/\sqrt{2}, \tag{7.62}$$

respectively in the $O(3)$ regime. Since the dot conjugation is the parity operation, they are like

$$-\dot{u}\dot{u}, \qquad -\dot{v}\dot{v}, \qquad -(\dot{u}\dot{v} + \dot{v}\dot{u})/\sqrt{2}. \tag{7.63}$$

In other words,

$$(uu\dot{)} = -\dot{u}\dot{u} \quad \text{and} \quad (vv\dot{)} = -\dot{v}\dot{v}. \tag{7.64}$$

We noticed a similar sign change in Eq. (7.57).

In order to construct the z component in this $O(3)$ space, let us first consider

$$f_z = \frac{1}{2}\left[(uv + vu) - (\dot{u}\dot{v} + \dot{v}\dot{u})\right], \qquad g_z = \frac{1}{2i}\left[(uv + vu) + (\dot{u}\dot{v} + \dot{v}\dot{u})\right]. \tag{7.65}$$

Here, f_z and g_z are respectively symmetric and anti-symmetric under the dot conjugation or the parity operation. These quantities are invariant under the boost along the z direction. They are also invariant under rotations around this axis, but they are not invariant under boost along or rotations around the x or y axis. They are different from the scalars given in Eq. (7.51).

Next, in order to construct the x and y components, we start with $f_\pm$ and $g_\pm$ as

$$f_+ = \frac{1}{\sqrt{2}}(uu - \dot{u}\dot{u}) \qquad f_- = \frac{1}{\sqrt{2}}(vv - \dot{v}\dot{v})$$

$$g_+ = \frac{1}{\sqrt{2}i}(uu + \dot{u}\dot{u}) \qquad g_- = \frac{1}{\sqrt{2}i}(vv + \dot{v}\dot{v}). \tag{7.66}$$

Then

$$f_x = \frac{1}{\sqrt{2}}(f_+ + f_-) = \frac{1}{2}\left[(uu + vv) - (\dot{u}\dot{u} + \dot{v}\dot{v})\right]$$

$$f_y = \frac{1}{\sqrt{2}i}(f_+ - f_-) = \frac{1}{2i}\left[(uu - vv) - (\dot{u}\dot{u} - \dot{v}\dot{v})\right], \tag{7.67}$$

and

$$g_x = \frac{1}{\sqrt{2}} (g_+ + g_-) = \frac{1}{2} [(uu + vv) + (\dot{u}\dot{u} + \dot{v}\dot{v})]$$

$$g_y = \frac{1}{\sqrt{2}i} (g_+ - g_-) = \frac{1}{2i} [(uu - vv) + (\dot{u}\dot{u} - \dot{v}\dot{v})]. \tag{7.68}$$

Here, f_x and f_y are symmetric under dot conjugation, while g_x and g_y are anti-symmetric.

Furthermore, f_z, f_x, and f_y of Eqs. (7.65) and (7.67) transform like a 3D vector. The same can be said for g_i of Eqs. (7.65) and (7.68). Thus, they can be grouped into the second-rank tensor

$$\begin{pmatrix} 0 & -f_z & -f_x & -f_y \\ f_z & 0 & -g_y & g_x \\ f_x & g_y & 0 & -g_z \\ f_y & -g_x & g_z & 0 \end{pmatrix}, \tag{7.69}$$

whose Lorentz-transformation properties are well known. The g_i components change their signs under space inversion, while the f_i components remain invariant. They are like the electric and magnetic fields respectively.

If the system is Lorentz-boosted, f_i and g_i can be computed from Table 7.2. We are now interested in the symmetry of photons by taking the massless limit. Thus, we keep only the terms which become larger for larger values of η. Thus,

$$f_x \rightarrow \frac{1}{2} (uu - \dot{v}\dot{v}), \qquad f_y \rightarrow \frac{1}{2i} (uu + \dot{v}\dot{v}),$$

$$g_x \rightarrow \frac{1}{2i} (uu + \dot{v}\dot{v}), \qquad g_y \rightarrow -\frac{1}{2} (uu - \dot{v}\dot{v}), \tag{7.70}$$

in the massless limit.

Then the tensor of Eq. (7.69) becomes

$$\begin{pmatrix} 0 & 0 & -E_x & -E_y \\ 0 & 0 & -B_y & B_x \\ E_x & B_y & 0 & 0 \\ E_y & -B_x & 0 & 0 \end{pmatrix}, \tag{7.71}$$

with

$$E_x \simeq \frac{1}{2} (uu - \dot{v}\dot{v}), \qquad E_y \simeq \frac{1}{2i} (uu + \dot{v}\dot{v}),$$

$$B_x = \frac{1}{2i} (uu + \dot{v}\dot{v}), \qquad B_y = -\frac{1}{2} (uu - \dot{v}\dot{v}). \tag{7.72}$$

The electric and magnetic field components are perpendicular to each other. Furthermore,

$$B_x = E_y, \qquad B_y = -E_x. \tag{7.73}$$

In order to address symmetry of photons, let us go back to Eq. (7.66). In the massless limit,

$$B_+ \simeq E_+ \simeq uu, \qquad B_- \simeq E_- \simeq \dot{v}\dot{v}. \tag{7.74}$$

The gauge transformations applicable to u and $\dot{v}$ are the 2×2 matrices

$$\begin{pmatrix} 1 & -\gamma \\ 0 & 1 \end{pmatrix} \quad \text{and} \quad \begin{pmatrix} 1 & 0 \\ \gamma & 1 \end{pmatrix}, \tag{7.75}$$

respectively. Both u and $\dot{v}$ are invariant under gauge transformations, while $\dot{u}$ and v are not.

The B_+ and E_+ are for the photon spin along the z direction, while B_- and E_- are for the opposite direction.

7.6.3   *Higher Spins*

Since Wigner's original book of 1931 (Wigner, 1931, 1959), the rotation group, without Lorentz transformations, has been extensively discussed in the literature (Condon and Shortley, 1979; Hamermesh, 1989). One of the main issues was how to construct the most general spin state from the two-component spinors for the spin-1/2 particle.

Since there are two states for the spin-1/2 particle, four states can be constructed from two spinors, leading to one state for the spin-0 state and three spin-1 states. With three spinors, it is possible to construct four spin-3/2 states and two spin-1/2 states, resulting in six states. This partition process is much more complicated (Feynman *et al.*, 1971; Hussar *et al.*, 1980) for the case of three spinors. Yet, this partition process is possible for all higher-spin states.

In the Lorentz-covariant world, there are four states for each spin-1/2 particle. With two spinors, we end up with 16 (4×4) states, and they are tabulated in Table 7.2. There should be 64 states for three spinors, and 256 states for four spinors. We now know how to Lorentz-boost those spinors. We also know that the transverse rotations become gauge transformations in the limit of zero-mass or infinite-η. It is thus possible to bundle all of them into the table given in Fig. 7.4.

In the relativistic regime, we are interested in photons and gravitons. As was noted in Subsecs. 7.6.1 and 7.6.2, the observable components are

	Spin 1/2	Spin 1	Higher Spin
Massive		Rotations	
Massless		Helicity Gauge Trans.	

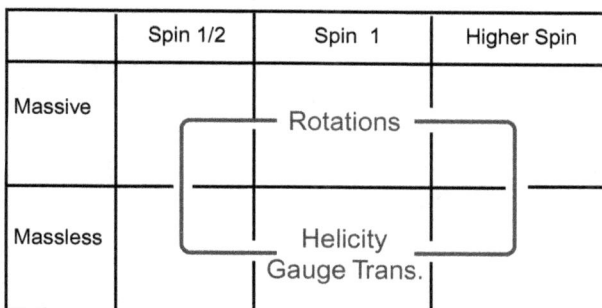

Fig. 7.4.   Unified picture of massive and massless particles. The gauge transformation is a Lorentz-boosted rotation matrix, and is applicable to all massless particles. It is possible to construct higher-spin states starting from the four states of the spin-1/2 particle in the Lorentz-covariant world.

invariant under gauge transformations. They are also the terms which become largest for large values of η.

We have seen in Subsec. 7.6.2, the photon state consists of uu and $\dot{v}\dot{v}$ for those whose spins are parallel and anti-parallel to the momentum respectively. Thus, for spin-2 gravitons, the states must be $uuuu$ and $\dot{v}\dot{v}\dot{v}\dot{v}$ respectively.

In his effort to understand photons and gravitons, Weinberg constructed his states for massless particles (Weinberg, 1964b), especially photons and gravitons (Weinberg, 1964c). He started with the conditions

$$N_1|\text{state}\rangle = 0 \quad \text{and} \quad N_2|\text{state}\rangle = 0, \tag{7.76}$$

where N_1 and N_2 are defined in Eq. (5.39). Since they are now known as the generators of gauge transformations, Weinberg's states are gauge-invariant states. Thus, uu and $\dot{v}\dot{v}$ are Weinberg's states for photons, and $uuuu$ are $\dot{v}\dot{v}\dot{v}\dot{v}$ are Weinberg's states for gravitons.

Chapter 8

Lorentz-covariant Harmonic Oscillators

Einstein and Bohr met occasionally, before and after 1927, to discuss physics. Einstein was interested in how things look to moving observers, while Bohr was interested in why the energy levels of the hydrogen atom are discrete. Then they must have talked about how the electron orbit of the hydrogen atom looks to a moving observer. There does not seem to be written records to indicate how they sketched the orbits.

However, it is not uncommon to see in the literature the description of the Lorentz deformation as described in Fig. 8.1. This figure became outdated in 1927. The electron orbit is now a standing wave. Thus, the question is how the standing wave appears when it is boosted along a given direction.

As is indicated in Fig. 8.1, the longitudinal component is affected while the transverse components remain unchanged. With this point in mind, we shall study harmonic oscillators. Since the wave equation for the three-dimensional oscillator is separable in the Cartesian coordinate system, it is sufficient to study the effect of the Lorentz boost only for the longitudinal component of the wave function.

8.1 Introduction

While Einstein's special relativity was and still is a powerful theory, Einstein did not like the probabilistic interpretation of wave functions. He had good reasons, and those reasons are well known. However, there is one important

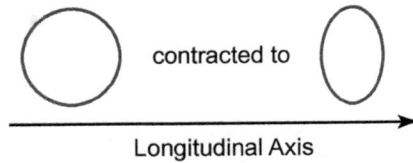

Fig. 8.1. Classical picture of Lorentz contraction of the electron orbit in the hydrogen atom. It is expected that the longitudinal component becomes contracted while the transverse components are not affected. In the first edition of his book published in 1987, 60 years after 1927, John S. Bell included this picture of the orbit viewed by a moving observer (Bell, 2004). While talking about quantum mechanics in his book, Bell overlooked the fact that the electron orbit in the hydrogen atom had been replaced by a standing wave in 1927. The question then is how standing waves look to moving observers.

reason not mentioned often in the literature. Is the concept of probability consistent with Einstein's Lorentz covariance?

According to his note on Einstein written in 1974, Heisenberg confessed that he could not accept Einstein's simultaneity even though he liked the mathematics of the Lorentz group. Heisenberg was saying that the concept of simultaneous measurement is not consistent with Einstein's Lorentz covariance.

Paul A. M. Dirac was more specific. In 1927 (Dirac, 1927), he studied the uncertainty relation between the time and energy variables and concluded that it does not allow quantum excitations along these variables. Thus, space and time cannot be linearly combined as in special relativity. In 1949 (Dirac, 1949), Dirac listed three possible ways to construct relativistic quantum mechanics, and pointed out the difficulties in all these approaches.

In 1971, Feynman with his students published a paper (Feynman *et al.*, 1971) in which they wrote down a Lorentz-invariant differential equation for an object consisting of two quarks (fundamental particles) bound together by a harmonic oscillator force. However, they ended up with oscillator wave functions not normalizable in the time separation variable. Their wave functions are not acceptable in quantum mechanics.

In this chapter, we translate Dirac's efforts and Feynman's efforts into the language of Wigner's little group for massive particles and construct Lorentz-covariant harmonic oscillators carrying a Lorentz covariant probability interpretation.

In Sec. 8.2, it is noted that Dirac wrote three important papers to address the problems of combining quantum mechanics with relativity, and that these papers can be synthesized to produce a wave function that can be Lorentz-boosted. In Sec. 8.3, we start with the Lorentz-invariant differential equation for two particles bound together by a harmonic oscillator force. This equation allows us to separate the internal coordinates for a standing wave and the external coordinate for a running wave.

In Sec. 8.4, we note that Wigner's little group is applicable to the internal coordinates, and then construct a representation based on the harmonic oscillator wave functions. In Secs. 8.5 and 8.6, we study transformation properties and Lorentz contractions of oscillator wave functions, respectively.

In Sec. 8.7, it is shown that the covariant oscillator formalism shares the same set of physical principles as Feynman diagrams. In Sec. 8.8, the role of the covariant oscillators is discussed for the quantum field theory of extended particles.

8.2 Dirac's Approach to Lorentz-covariant Wave Functions

Paul A. M. Dirac spent many years trying to construct a quantum mechanics consistent with special relativity. Dirac published the following three important papers for this purpose.

1. In 1927, Dirac considered the uncertainty between the time and energy variables, and noted that there are no excitations along these coordinates [Dirac (1927)]. Thus, he concluded that the time-energy uncertainty relation is a *c-number* relation. He pointed out the problem with the Heisenberg uncertainty relation defined only for position and momentum. What happens when the time and space variables are linearly mixed under the Lorentz boost?
2. In 1945 (Dirac, 1945b), Dirac wrote down the Gaussian form

$$\exp\left[-\frac{1}{2}\left(x^2 + y^2 + z^2 + t^2\right)\right], \tag{8.1}$$

and said this could be the starting point for building harmonic oscillator wave functions that can be Lorentz-boosted. However, he did not explain why the exponent was not in the Lorentz-invariant form of $\left(x^2 + y^2 + z^2 - t^2\right)$. Furthermore, he did not explain why the wave function has to vanish for remote past and remote future with large values of $|t|$.

3. In 1949 (Dirac, 1949), Dirac mentions three possible ways of making quantum mechanics compatible with relativity, but he ends up with mentioning the difficulties.

However, there are soft spots in Dirac's papers.

1. Dirac did not use the results he obtained in his earlier papers. In his 1945 paper on harmonic oscillators, he forgot what he said in his 1927 paper where the time variable should be treated differently.
2. In his 1945 paper, Dirac should have noted the difference between time and the time difference. For instance, the ages of son and father are increased at the same rate. However, their age difference does not change over the time. The Bohr radius is the spatial separation between the proton and electron. Likewise, the time variable in his Gaussian form should be a sign difference.
3. In his 1949 paper, he introduced his *light cone* coordinate system for Lorentz boosts. In this system, the Lorentz boost is carried out by a diagonal matrix. However, he never attempted to Lorentz-boost the Gaussian form he introduced in his earlier paper of 1945.
4. Again in his 1949 paper, Dirac shows that the extension of Heisenberg's uncertainty relations leads to the Lie algebra of the Poincaré group or inhomogeneous Lorentz group. He was aware of Wigner's 1939 paper on this subject (Wigner, 1939). However, he did not make any attempts to exploit the contents of this paper with his earlier works, namely the space–time asymmetry he noted in his 1927 paper (Dirac, 1927) and the oscillator wave function he wrote down in 1945 (Dirac, 1945b).
5. Dirac's papers are like poems, but they do not contain figures or diagrams, as was pointed out in Table 1.1.

In this chapter, we remove these soft spots using the language of Wigner's little groups. As in Fig. 8.2, we combine Dirac's 1927 and 1945 papers into a circle. This figure also gives an illustration of the Lorentz boost in his light-cone coordinate system which Dirac introduced in 1945. During the boost, $t^2 - z^2$ remains constant, but this quantity can also be written as $(t + z)(t - z)$. Thus, it is an area-preserving squeeze transformation. Indeed, according to Dirac, the Lorentz boost is a squeeze transformation.

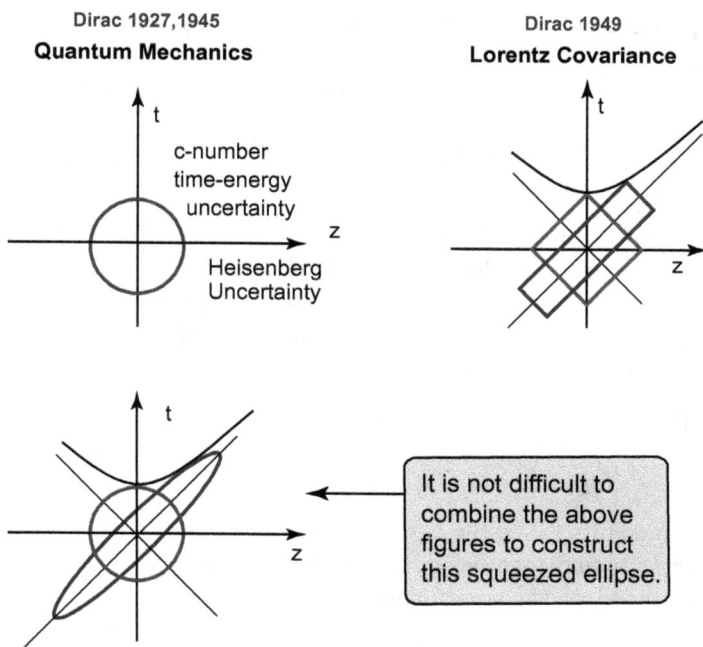

Dirac 1927,1945
Quantum Mechanics

Dirac 1949
Lorentz Covariance

It is not difficult to combine the above figures to construct this squeezed ellipse.

Fig. 8.2. Dirac's Lorentz-boosted oscillator wave function. His c-number time-energy uncertainty relation (Dirac, 1927) and his oscillator wave function (Dirac, 1945b) can be combined into a circle. His light-cone coordinate system (Dirac, 1949) can be illustrated as a squeeze of the square into a rectangle. This is the point of this chapter.

8.3 Running Waves and Standing Waves

We are quite familiar with the Klein–Gordon equation

$$\left[-\left(\frac{\partial}{\partial x} \right)^2 + m^2 \right] \phi(x) = 0, \tag{8.2}$$

where

$$\left(\frac{\partial}{\partial x} \right)^2 = \left(\frac{\partial}{\partial x} \right)^2 + \left(\frac{\partial}{\partial y} \right)^2 + \left(\frac{\partial}{\partial z} \right)^2 - \left(\frac{\partial}{\partial t} \right)^2 \tag{8.3}$$

for a single particle, where x is a four vector and thus

$$x^2 = x^2 + y^2 + z^2 - t^2. \tag{8.4}$$

The x, y and z on the right-hand side of this equation are coordinate variables.

For two particles with masses m_a and m_b, the equation becomes

$$\left[-\left(\frac{\partial}{\partial x_a}\right)^2 - \left(\frac{\partial}{\partial x_b}\right)^2 + m_a^2 + m_b^2 \right] \phi(x_a)\,\phi(x_b) = 0, \qquad (8.5)$$

where x_a and x_b are their space–time four vectors. The physics of this system for two free particles is also well known.

The question is what happens when $(m_a^2 + m_b^2)$ is replaced by a term containing $(x_a - x_b)^2$. Indeed, in 1971, Feynman, Kislinger, and Ravndal wrote down the equation (Feynman *et al.*, 1971)

$$\left\{ -2\left[\left(\frac{\partial}{\partial x_a}\right)^2 + \left(\frac{\partial}{\partial x_b}\right)^2 \right] + \frac{1}{16}(x_a - x_b)^2 + m_0^2 \right\} \phi(x_a, x_b) = 0,$$
$$(8.6)$$

for the bound state called the *hadron* consisting of two constituent particles called *quarks* bound together in a harmonic oscillator potential.

Let us introduce the coordinates

$$X = \frac{x_a + x_b}{2}, \quad \text{and} \quad x = \frac{x_a - x_b}{2\sqrt{2}}. \qquad (8.7)$$

The X coordinate is for the space–time specification of the hadron, while the x variable measures the relative space–time separation between the quarks. In terms of these variables, Eq. (8.6) can be written as

$$\left\{ -\left(\frac{\partial}{\partial X}\right)^2 + m_0^2 - \frac{1}{2}\left[\left(\frac{\partial}{\partial x}\right)^2 + x^2 \right] \right\} \phi(X, x) = 0. \qquad (8.8)$$

This equation is separable in the X and x variables, and can be separated by using the equation:

$$\phi(X, x) = f(X)\,\psi(x), \qquad (8.9)$$

where $f(X)$ and $\psi(x)$ satisfy the following differential equations respectively

$$\left\{ -\left(\frac{\partial}{\partial X}\right)^2 + m_0^2 + (\lambda + 1) \right\} f(X) = 0, \qquad (8.10)$$

and

$$\frac{1}{2}\left[\left(-\frac{\partial}{\partial x}\right)^2 + x^2 \right] \psi(x) = (\lambda + 1)\psi(x). \qquad (8.11)$$

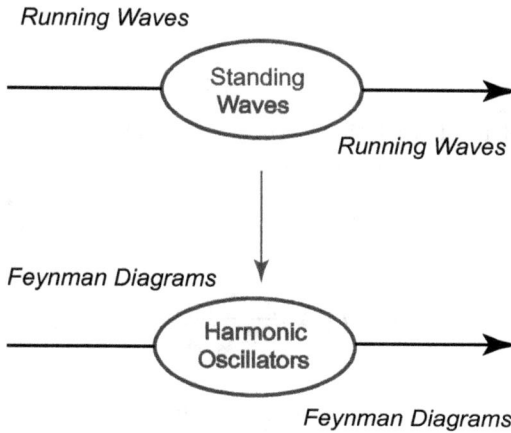

Running Waves

Standing
Waves

Running Waves

Feynman Diagrams

Harmonic
Oscillators

Feynman Diagrams

Fig. 8.3. Feynman's suggestion for combining quantum mechanics with special relativity. Feynman diagrams work for running waves, and they provide a satisfactory interpretation of scattering states in the Lorentz-covariant world. For standing waves trapped inside a bound state, (Feynman *et al.*, 1971) suggested harmonic oscillators as the first step.

The differential equation of Eq. (8.10) is a Klein–Gordon equation, and its solution is well known. It takes the form

$$f(X) = \exp\left(\pm ip \cdot X\right) \tag{8.12}$$

with

$$-p^2 = m_0^2 + (\lambda + 1), \tag{8.13}$$

where p is the four-momentum of the hadron. $-p^2$ is, of course, the $(mass)^2$ of the hadron and is numerically constrained to take the values allowed by Eq. (8.13). The separation constant λ is determined from the solutions of the harmonic oscillator equation of Eq. (8.11).

Indeed, the wave function of Eq. (8.9) is for the hadron moving with the four-momentum p_μ with the internal structure dictated by the oscillator equation, as is described in Fig. 8.3. Wigner's little group is applicable to the internal space–time symmetry dictated by the oscillator equation of Eq. (8.11) (Kim *et al.*, 1979a; Kim and Noz, 1986).

The space–time transformation of the total wave function of Eq. (8.9) is generated by the following 10 generators of the Poincaré group.

The operators

$$P_\mu = -i \frac{\partial}{\partial X^\mu} \qquad (8.14)$$

generate space–time translations. Lorentz transformations, which include boosts and rotations, are generated by

$$M_{\mu\nu} = L^*_{\mu\nu} + L_{\mu\nu} \qquad (8.15)$$

where

$$L^*_{\mu\nu} = -i \left(X_\mu \frac{\partial}{\partial X^\nu} - X_\nu \frac{\partial}{\partial X^\mu} \right),$$

$$L_{\mu\nu} = -i \left(x_\mu \frac{\partial}{\partial x^\nu} - x_\nu \frac{\partial}{\partial x^\mu} \right).$$

The translation operators P_μ, act only on the hadronic coordinate, and do not affect the internal coordinate. The operators $L^*_{\mu\nu}$ and $L_{\mu\nu}$ Lorentz-transform the hadronic and internal coordinates, respectively. The above 10 generators satisfy the commutation relations for the Poincaré group (Dirac, 1949).

In order to consider irreducible representations of the Poincaré group, we have to construct wave functions which are diagonal in the invariant Casimir operators of the group, which commute with all the generators of Eqs. (8.14) and (8.15). The Casimir operators in this case are

$$P^\mu P_\mu, \quad \text{and} \quad W^\mu W_\mu, \qquad (8.16)$$

where

$$W_\mu = \epsilon_{\mu\nu\alpha\beta} P^\nu M^{\alpha\beta}. \qquad (8.17)$$

The eigenvalues of the above $-P^2$ and $-W^2$ represent respectively the mass and spin of the hadron.

The algebra of these generators becomes much simpler when the hadronic momentum is constant, as in the case of Wigner's little group. While translation generators can be dropped from the algebra, the operator P^ν can be replaced by the four-vector

$$p = (0, 0, p, E) \qquad (8.18)$$

for the hadron momentum moving in the z direction. As a consequence, the eigenvalues of the Casimir operators become $m^2 = (mass)^2$ and $m^2 \ell(\ell+1)$, where ℓ is the total angular momentum of oscillator. These eigenvalues are invariant under Poincaré or Lorentz transformations.

8.4 Little Groups for Relativistic Extended Particles

The harmonic oscillator equation of Eq. (8.11) is invariant under Lorentz transformations. For instance, if the system is boosted along the z direction:

$$\begin{pmatrix} z' \\ t' \end{pmatrix} = \begin{pmatrix} \cosh\eta & \sinh\eta \\ \sinh\eta & \cosh\eta \end{pmatrix} \begin{pmatrix} z \\ t \end{pmatrix}, \tag{8.19}$$

the differential equation takes the same form in the new coordinate variables. Thus, the solution also takes the previous form. With this point in mind, we can now study the solution of the differential equation in the Lorentz frame where the hadron is at rest. Let us spell out the oscillator equation.

$$\frac{1}{2}\left\{\left[x^2 + y^2 + z^2 - \left(\frac{\partial}{\partial x}\right)^2 - \left(\frac{\partial}{\partial y}\right)^2 - \left(\frac{\partial}{\partial z}\right)^2\right] \right. $$
$$\left. - \left[t^2 - \left(\frac{\partial}{\partial t}\right)^2\right]\right\}\psi(x) = (\lambda + 1)\psi(x). \tag{8.20}$$

According to Dirac's c-number time-energy uncertainty relation, the time component of the solution should be always in the ground state, and thus the solution takes the form

$$\psi(x) = \varphi(x, y, z)\left[\left(\frac{1}{\pi}\right)^{1/4} e^{-t^2/2}\right], \tag{8.21}$$

with

$$\frac{1}{2}\left[x^2 + y^2 + z^2 - \left(\frac{\partial}{\partial x}\right)^2 - \left(\frac{\partial}{\partial y}\right)^2 - \left(\frac{\partial}{\partial z}\right)^2\right]\varphi(z, x, y)$$

$$= \left(\lambda + \frac{3}{2}\right)\varphi(z, x, y). \tag{8.22}$$

This equation is very familiar to us from textbooks. However, the equation carried the following additional interpretations.

1. The Cartesian variables z, x, and y are internal coordinate variables for the hadron.
2. This equation is separable in both the spherical and Cartesian coordinate system. For the discussion of the Poincaré symmetry, we need the spherical coordinate system to construct the representation diagonal in the Casimir operators where the eigenvalue ℓ is needed.

3. When the system is boosted along the z direction, the transverse x and y are not affected, and they can be separated out from the differential equation of Eq. (8.20).
4. The spherical solutions can be written in terms of the linear combinations of the Cartesian solutions.

The solution in the spherical coordinate system should take the form

$$\varphi(r, \theta, \phi) = R_{n\ell m}(r) Y_{\ell m}(\theta, \phi), \tag{8.23}$$

where $Y_{\ell m}(\theta, \phi)$ are the spherical harmonics. The radial function $R_{n\ell m}(r)$ is well defined, but its explicit form is not readily available in the literature. It should take the form

$$R_{n\ell m}(r) = r^m g_{n\ell}(r) e^{-r^2/2}, \tag{8.24}$$

with

$$g_{n\ell} = \sum_k a_{2k} r^{2k}, \tag{8.25}$$

where

$$\frac{a_{2k+2}}{a_{2k}} = \frac{2(\lambda - \ell - 2k)}{\ell(\ell+1) - (\ell+2k+3)(\ell+2k+2)}. \tag{8.26}$$

For large values of k, this ratio becomes $1/k$, which is like the expansion of the exponential $\exp(r^2)$, leading the radial function of Eq. (8.24) to increase as $\exp(r^2/2)$. Thus, the series has to be truncated with

$$\lambda = 2k + \ell. \tag{8.27}$$

The first term a_0 in the series is determined by the normalization condition

$$\int_0^\infty [r R_{n\ell m}(r)]^2 \, dr = 1. \tag{8.28}$$

The increases in ℓ and n are called the orbital and radial excitations in the literature.

If the system is Lorentz-boosted along the z direction according to Eq. (8.19), the Lorentz-invariant differential equation of Eq. (8.11) remains invariant. The z and t variables in Eq. (8.20) become z' and t', respectively, and the wave function becomes modified accordingly. The important point is that the eigenvalue λ remains invariant.

8.5 Transformation Properties of the Covariant Oscillator Wave Functions

We are now interested in what happens to the oscillator wave functions when they are Lorentz-boosted. Since the x and y coordinates are not affected, we drop their terms from the differential equation from Eq. (8.11), and consider the equation

$$\frac{1}{2}\left\{\left[z^2 - \left(\frac{\partial}{\partial z}\right)^2\right] - \left[t^2 - \left(\frac{\partial}{\partial t}\right)^2\right]\right\}\psi(z,t) = \lambda\psi(z,t). \quad (8.29)$$

The solution of this equation should take the form

$$\psi_0^n(z,t) = \chi_n(z)\chi_0(t) = \left[\frac{1}{\pi 2^n n!}\right]^{1/2} H_n(z)\exp\left[-\frac{1}{2}\left(z^2 + t^2\right)\right], \quad (8.30)$$

where $H_n(z)$ is the Hermite polynomial of order n, and $\chi_n(z)$ is the oscillator wave function in the nth excited state:

$$\chi_n(z) = \left[\frac{1}{\sqrt{\pi}2^n n!}\right]^{1/2} H_n(z)\exp\left(-\frac{1}{2}z^2\right). \quad (8.31)$$

The differential equation of Eq. (8.29) is invariant under the Lorentz boost along the z direction, and is invariant under the replacements of z and t by z' and t' respectively, where

$$\begin{pmatrix} z' \\ t' \end{pmatrix} = \begin{pmatrix} \cosh\eta & -\sinh\eta \\ -\sinh\eta & \cosh\eta \end{pmatrix} \begin{pmatrix} z \\ t \end{pmatrix}. \quad (8.32)$$

This is the inverse of the transformation given in Eq. (8.19). This difference is of course the way in which the arguments of the function and coordinates are transformed in the opposite directions.

We thus achieve the Lorentz boost of the wave function by writing

$$\psi_\eta^n(z,t) = \psi_0^n(z',t') = \left[\frac{1}{\pi 2^n n!}\right]^{1/2} H_n(z')\exp\left[-\frac{1}{2}\left(z'^2 + t'^2\right)\right]. \quad (8.33)$$

For $n = 0$, the wave function in the Gaussian form is

$$\psi_\eta^0(z,t) = \left(\frac{1}{\pi}\right)^{1/2}\exp\left\{-\frac{1}{4}\left[e^{-2\eta}(z+t)^2 + e^{2\eta}(z-t)^2\right]\right\}. \quad (8.34)$$

This Gaussian form corresponds to the circle and ellipse in Fig. 8.2, and illustrates how the wave function is deformed when Lorentz-boosted.

We are now interested in writing $\psi_\eta^n(z,t)$ of Eq. (8.33) as a series expansion in $\chi_k(z)$ and $\chi_k(t)$:

$$\psi_\eta^n(z,t) = \sum_{k',k} A_{k',k}\chi_{k'}(z)\chi_k(t). \qquad (8.35)$$

We then have to evaluate the coefficient $A_{k',k}$.

First of all, we can write the differential equation of Eq. (8.29) as

$$\frac{1}{2}\left\{ \left(z^2 - \frac{\partial^2}{\partial z^2} \right) - \left(t^2 - \frac{\partial^2}{\partial t^2} \right) \right\} \chi_{k'}(z)\chi_k(t) = (k' - k)\chi_n(z)\chi_m(t). \qquad (8.36)$$

Since this equation is Lorentz-invariant, $(k' - k)$ is Lorentz-invariant, and $k' - k = n$. Hence, the series takes the form

$$\psi_\eta^n(z,t) = \sum_k A_k(n)\chi_{k+n}(z)\chi_k(t). \qquad (8.37)$$

This is a sum over a single index k, with

$$\sum_k |A_k(n)|^2 = 1. \qquad (8.38)$$

This coefficient is

$$A_k(n) = \int \chi_{k+n}(z)\chi_k(t)\chi_n(z')\chi_0(t')\, dz\, dt. \qquad (8.39)$$

In order to carry out this calculation, let us use the generating function of the Hermite polynomials (Magnus *et al.*, 1966):

$$G(r,z) = \exp\left(-r^2 + 2rz\right) = \sum_m \frac{r^m}{m!}H_m(z), \qquad (8.40)$$

and evaluate the integral

$$I = \int G(r,x)G(s,y)G(r',x')\exp\left\{ -\left(\frac{x^2 + y^2 + x'^2 + y'^2}{2} \right) \right\}\, dx\, dy. \qquad (8.41)$$

The integrand becomes one exponential function, and its exponent is quadratic in x and y. This quadratic form can be diagonalized, and the integral can be evaluated (Kim *et al.*, 1979b; Başkal *et al.*, 2016).

The result is

$$I = \left[\frac{\pi}{\cosh \eta}\right] \exp\left(2rs \tanh \eta\right) \exp\left(\frac{2rr'}{\cosh \eta}\right). \tag{8.42}$$

We can now Taylor-expand this expression and choose the coefficients of r^{n+k}, s^k, r'^n for $H_{(n+k)}(z), H_k(t)$, and $H_n(z')$ respectively. The result is

$$A_{n;k} = \left(\frac{1}{\cosh \eta}\right)^{(n+1)} \left[\frac{(n+k)!}{n!k!}\right]^{1/2} (\tanh \eta)^k. \tag{8.43}$$

Thus, the series becomes

$$\chi_n(z')\chi_0(t') = \left(\frac{1}{\cosh \eta}\right)^{(n+1)} \sum_k \left[\frac{(n+k)!}{n!k!}\right]^{1/2} (\tanh \eta)^k \chi_{k+n}(z)\chi_k(t). \tag{8.44}$$

The result is

$$\psi_\eta^n(z,t) = \left(\frac{1}{\cosh \eta}\right)^{(n+1)} \sum_k \left[\frac{(n+k)!}{n!k!}\right]^{1/2} (\tanh \eta)^k \chi_{n+k}(z)\chi_k(t), \tag{8.45}$$

where $\chi_n(z)$ is the nth excited state oscillator wave function which takes the familiar form

$$\chi_n(z) = \left[\frac{1}{\sqrt{\pi} 2^n n!}\right]^{1/2} H_n(z) \exp\left(-\frac{z^2}{2}\right). \tag{8.46}$$

If $n = 0$, this formula becomes simplified to

$$\psi_\eta^0(z,t) = \left(\frac{1}{\cosh \eta}\right)^{1/2} \sum_k (\tanh \eta)^k \chi_k(z)\chi_k(t). \tag{8.47}$$

8.6 Lorentz Contraction of Harmonic Oscillators

Let us now consider the overlap of the wave function $\psi_\eta^n(z,t)$ with $\eta = 0$, as indicated in Fig. 8.4. We are interested in the integral

$$\int \left(\psi_\eta^{n'}(z,t)\right)^* \psi_0^n(z,t) \, dz \, dt, \tag{8.48}$$

where $\psi_0^n(z,t)$, given in Eq. (8.30), can be written as

$$\psi_0^n = \chi_n(z)\chi_0(t). \tag{8.49}$$

Then the evaluation of this integral is straightforward from Eq. (8.36), and the result is (Ruiz, 1974)

$$\left(\psi_\eta^{n'}, \psi_0^n\right) = \left(\frac{1}{\cosh \eta}\right)^{(n+1)} \delta_{nn'}. \tag{8.50}$$

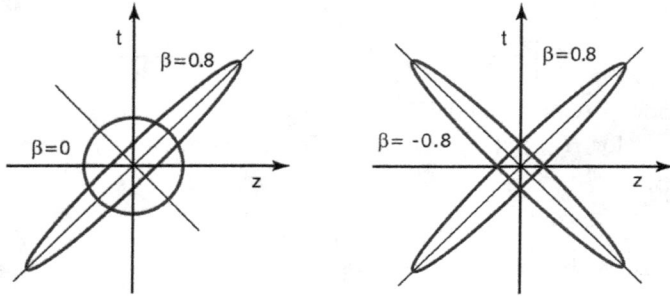

Fig. 8.4. Two overlapping wave functions. The circle is for the hadron at rest. If the hadron moves, the circle becomes squeezed into an ellipse. The wave functions moving in opposite directions are also shown. The ellipse is for $\beta = \pm 0.8$ (Kim and Noz, 1986). The area of overlaps corresponds to their inner product.

This means that the orthogonality relation is preserved between two wave functions in two different frames.

If $n = n'$, the inner product between two wave functions leads to the contraction given on the right-hand side of Eq. (8.50). In terms of the velocity parameter $\beta = v/c$, where v is the hadronic velocity,

$$\frac{1}{\cosh \eta} = \sqrt{1 - \beta^2}. \tag{8.51}$$

This expression is more familiar to us, and the right-hand side of Eq. (8.50) is

$$\left(\sqrt{1 - \beta^2}\right)^{(n+1)}. \tag{8.52}$$

For the ground-state wave function with $n = 0$, it is like the Lorentz contraction of a rigid body. For the first excited state, it is like an additional rod. This is not surprising in view of the fact that the excited states are obtained through application of the step-up operator. The nth excited state $|n >$ can be written as

$$\frac{1}{\sqrt{n!}} \left(a^\dagger\right)^n |0 >. \tag{8.53}$$

The additional contraction factor $\sqrt{1 - \beta^2}$ comes from the step-up operator.

If the value of η in one of the wave functions is replaced by the non-zero value η', $\cosh \eta$ in Eq. (8.51) should become $\cosh(\eta - \eta')$. Of particular interest is the case with $\eta' = -\eta$, as shown in Fig. 8.4. In this case, this

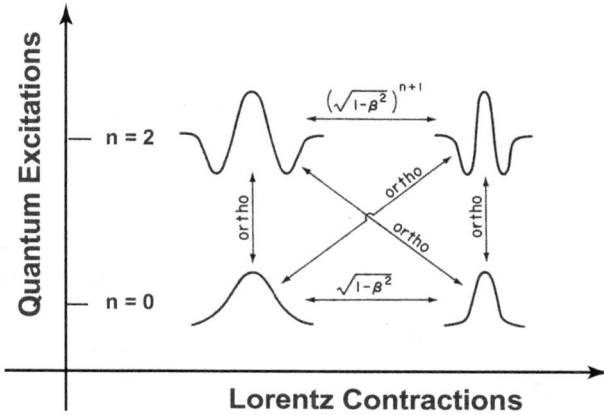

Fig. 8.5. Orthogonality relations for covariant oscillator wave functions. The harmonic oscillator can be excited and can also be Lorentz boosted. The orthogonality relations remain invariant under Lorentz boosts, and their inner products have the Lorentz-contraction property given in this figure (Ruiz, 1974; Kim and Noz, 1986).

is an overlap of two wave functions moving in the opposite directions, and the contraction factor is

$$\left(\frac{1}{\cosh(2\eta)}\right)^{(n+1)},\tag{8.54}$$

which, in the language of β, becomes

$$\left(\frac{1-\beta^2}{1+\beta^2}\right)^{(n+1)}.\tag{8.55}$$

Based on these orthogonality and contraction properties, it appears possible to give a quantum probability interpretation to the covariant harmonic oscillator in the Lorentz-covariant world as shown in Fig. 8.5.

On the other hand, these wave functions contain the time separation variable between the constituent particles. What is the meaning of the distribution in terms of this variable, which is thoroughly hidden in the present form of quantum mechanics? In order to clarify this issue, let us examine the concept of Feynman's rest of the universe in Chapter 10.

8.7 Physical Principles in Quantum Field Theory and in Covariant Oscillator Formalism

The formalism of Lorentz-covariant harmonic oscillators is shown to be consistent with quantum mechanics and special relativity. However, this formalism is applicable only to localized standing waves for hadrons. How about scattering states? The present form of quantum field theory provides a satisfactory answer to this problem. The tools for scattering problems are called *Feynman diagrams*. In Feynman diagrams, all the particles are free in remote past and remote future.

Thus, Feynman diagrams are not applicable to bound states but the study of harmonic oscillators could bring more fruitful results. Indeed, it was Feynman who said this in his invited talk at the April meeting of the American Physical Society held in 1970. This meeting took place at the Shoreham Hotel in Washington, DC, USA. Feynman published this talk with his students in 1971 (Feynman *et al.*, 1971). Indeed, it was Feynman who encouraged the authors of this book to study harmonic oscillators that can be Lorentz-boosted. They published their first paper in 1973 (Kim and Noz, 1973).

The physics of Feynman diagrams are well known. The question is whether the covariant harmonic oscillators share the same set of rules with Feynman diagrams (Han *et al.*, 1981). In this chapter, we have synthesized three of Dirac's papers to construct the covariant harmonic oscillators. The table in Fig. 8.6 tells us where this formalism stands in the history of

Photograph by A. John Coleman, courtesy AIP Emilio Segrè Visual Archives, Physics Today Collection.

Scattering	Bound States	Space/Time
COMET	PLANET	
Newton		
Rutherford	Bohr	GALILEI
WAVE-PARTICLE DUALITY Heisenberg, Schrödinger		
Running Waves Feynman	Standing Waves Dirac***	EINSTEIN
Put them into ONE BOX		

*** Synthesize three papers written by Dirac.

Fig. 8.6.   Scattering and bound states. Feynman is responsible for developing Feynman diagrams. By synthesizing Dirac's three papers, it is possible to obtain the covariant harmonic oscillator formalism. The question is how to construct one theory combining both.

physics. According to this figure, theories for scattering states and bound states were developed separately in the history of physics. It took major revolutionary steps to unify both theories into one. As is specified in the table, quantum field theory primarily deals with scattering problems, while the oscillators are for bound state problems.

Indeed, Feynman diagrams tell us about the real world, especially in quantum electrodynamics with spectacular numerical agreements with the experimental values. As will be seen in Chapter 9, the covariant harmonic oscillators give answers to the basic phenomena in high-energy physics. It would indeed be a revolution if there is one theory which will produce both Feynman diagrams and covariant oscillations.

We are not able to produce this unified theory. However, this does not prevent us from questioning whether Feynman diagrams and covariant oscillators share the same set of physical principles (Han *et al.*, 1981).

Question 1. Can we construct a field theory of extended hadrons where the standard Feynman rules are applicable to hadrons with free-particle asymptotic states, while the oscillator-like formalism is used to describe the internal motion of confined quarks?

Question 2. Although quantum field theory and the covariant harmonic oscillator formalism take different mathematical forms, it is possible that they are based on the same physical principles. If so, what are the physical principles underneath these two different-looking theories?

As for the first question, there are already several approaches based on this idea (Sogami, 1973; Kim, 1976; Karr, 1976; Capri and Chiang, 1976, 1977). The purpose of this section is to discuss the second question. For this purpose, we note first that both formalisms are covariant under Lorentz transformations. While the Feynman propagator and the wave function are the primary mathematical devices in field theory and the oscillator formalism, respectively, both quantities contain time-energy uncertainty in addition to Heisenberg's position-momentum uncertainty relation. We note further that both formalisms start with relativistic wave equations with negative energy spectra, and that both have their own respective ways of taking care of them. We shall discuss these points more systematically in the following subsections.

8.7.1　*Physical Principles in Quantum Field Theory*

Quantum field theory starts with the physical principles applicable to non-relativistic quantum mechanics including of course Heisenberg's position-momentum uncertainty relation. The question is then what additional physical principles are used in this theory.

First of all, field theory equations are relativistic. Thus, Lorentz covariance is one of the basic additional ingredients. Next, let us look into the question of causality. The present form of field theory starts with the causal commutator

$$[\phi(x), \phi(x')] = i\Delta(x - x') \tag{8.56}$$

where $\phi(x)$ is the field operator satisfying the Klein–Gordon equation. x is a space–time four-vector.

This commutation relation corresponds to Heisenberg's uncertainty relation when $x = x'$. The causal Green's function $\Delta(x - x')$ vanishes outside the light cone where

$$(t - t')^2 < (x - x')^2 - (y - y')^2 - (z - z')^2. \tag{8.57}$$

This means that the signal connecting two space–time points cannot propagate faster than light. In carrying out field theory calculations, however, we use more often the Feynman propagation function whose mathematical form is

$$\Delta_D(x - x') = \left(\frac{1}{2\pi}\right)^4 \int \frac{\exp\left(-ik \cdot (x - x')\right)}{k^2 - m^2 + i\epsilon} \, d^4k. \tag{8.58}$$

This function does not vanish outside the light cone. The reason for this is that the particle is no longer on the mass shell, and this unobservable particle does not respect causality. We are then naturally led to ask what additional physical laws are needed to explain this causality violation.

The derivation of the Feynman propagation function requires time and normal orderings which require *trading* ground-state energies with vacuum. In order to see the basic physics involved in this procedure more clearly, we now resort to the so-called *old-fashioned* field theory. Since all the basic physical concepts in the present covariant form of quantum field theory are contained in the *old-fashioned* field theory, it is not uncommon to refer to this earlier formalism in order to find explanations for what we do in the modern version of field theory (Weinberg, 1966).

For the purpose of finding the physical principles which led to the microscopic violation of causality and to the concept of *virtual* off-mass-shell particles, let us look at Chapter 4, Section 14 of Heitler's book (Heitler, 1984). The *old-fashioned* derivation of the Klein–Nishina formula is based on second-order time-dependent perturbation theory involving two integrations over the time variable. The second-time integration is done correctly from the interval 0 to t. However, the first-time integration contains a causality violation for the period allowed by the *c-number* time-energy uncertainty relation (Dirac, 1927). This introduction of time-energy uncertainty leads to the concept of virtual particles.

In the modern version of quantum field theory, the time-energy uncertainty manifests itself in the off-mass-shell particles contained in Feynman propagators. Although there is no precise definition of bound-state conditions in field theory, it is by now a widely accepted view that particles in bound states, which are affected by interactions, are not on their mass shells. In the following subsection, we shall see how this appears in the covariant harmonic oscillator model which deals only with quarks permanently bound inside a relativistic hadron.

8.7.2 *Physical Principles in the Covariant Oscillator Formalism*

The purpose of this subsection is to demonstrate that the covariant harmonic oscillator formalism employs the same physical principles as in quantum field theory, namely the Lorentz covariance and the c-number time-energy uncertainty, in addition to Heisenberg's position–momentum uncertainty relation.

First, the concept of c-number time–energy uncertainty relation is one of the basic additional ingredients in the covariant oscillator formalism. For a hadron at rest consisting of two quarks bound together by a harmonic oscillator potential of unit strength, the ground-state wave function takes the Gaussian form

$$\psi(z,t) = \frac{1}{\sqrt{\pi}} \exp\left[-\left(\frac{z^2 + t^2}{2}\right)\right], \qquad (8.59)$$

where z and t are the longitudinal and time-like coordinate separations between the quarks. The existence of the ground-state wave function in the time separation variable t (without excitations) is consistent with Dirac's c-number time–energy uncertainty relation.

Second, the oscillator formalism, as in the case of quantum field theory, violates microscopic causality. The wave function given in Eq. (8.59) does not vanish for the space-like separation:

$$t^2 < z^2. \tag{8.60}$$

This causality violation is not unlike that needed in the old-fashioned derivation of the Klein–Nishina formula in second-order time-dependent perturbation theory (Heitler, 1984).

Let us start with the Klein–Gordon equation of Eq. (8.2) for two free quarks, whose space–time coordinates are x_a and x_b, respectively. The wave function is separable in these coordinate variables. Then (Feynman *et al.*, 1971) coupled them with an oscillator potential with the differential equation of Eq. (8.6). This function is not separable in the x_a and x_b variables.

However, it was possible to introduce the X and x variables as in Eq. (8.7). Then the differential equation becomes separable in these variables. The equation for the X variable is the Klein–Gordon equation for the hadron with the momentum–energy four-vector for a free particle.

The x variable is for the space–time separation for the two quarks. During this separation process, the mass terms m_a^2 and m_b^2 become absorbed into the separation constant. This means that $(m_a^2 + m_b^2)$ is constant, but the m_a^2 or m_b^2 is no longer separately constant (Han and Kim, 1980). The mass of the quark inside the hadron is not on its mass shell (Han *et al.*, 1981).

8.8 Further Field Theoretic Concepts in the Covariant Oscillator Formalism

It is important to note that the Klein–Gordon equation of Eq. (8.10) and its solution $f(X)$ given in Eq. (8.12) are also important elements of the covariant harmonic oscillator formalism, as illustrated in Fig. 8.3.

It is possible to second quantize this Klein–Gordon wave and construct a field theory, scattering matrices, and Feynman diagrams for hadrons, while the harmonic oscillator wave functions describe the internal structure of the hadron. Indeed, many authors attempted to construct the field theory of hadrons with the oscillator wave functions (Kim, 1976; Karr, 1976; Capri and Chiang, 1976, 1977). This line of research is a *nonlocal* or *bilocal* field theory (Sogami, 1973; Mita, 1978).

When these papers were published, it was not known that the covariant harmonic oscillator formalism can serve as a representation of Wigner's little group dictating the internal space–time symmetry of a localized particle. Unknown also was that the Dirac spinors serve as the representation space for Wigner's little group. With these new tools, it is worthwhile to study those earlier papers on quantum field theory and Feynman diagrams for extended particles.

Chapter 9

Quarks and Partons

In Chapter 8, we constructed a set of harmonic oscillator wave functions that can be Lorentz-boosted. With this mathematical tool, we can study the main issue raised in Chapter 1 and again in Chapter 8. The main issue is what Einstein and Bohr forgot to address: how would the hydrogen atom look to a moving observer or how the hydrogen atom looks when it moves with a speed comparable with that of light. The purpose of this chapter is to provide a quantitative approach to this problem.

9.1 Introduction

We still do not have hydrogen atoms moving with relativistic speed. However, it is possible to study moving protons with speed close to that of light. Like the hydrogen atom, the proton is a bound state of more fundamental particles based on the same principles of quantum mechanics (Gell-Mann, 1964).

The proton is not a point particle. Like a hydrogen atom, it has a localized probability distribution. The question is how the distribution appears to moving observers. Indeed, thanks to the development of high-energy physics of the past century, the effects of these deformations have been measured, and many experimental plots are available in terms of the proton speed (Hofstadter, 1956).

Another question is what would happen to the probability distribution when the proton speed becomes very close to that of light. For this case, Feynman observed that the same proton looks like a collection of an infinite number of partons (Feynman, 1969a,b). Can the oscillator wave function answer this question?

In Sec. 9.2, it is noted that there has been an evolution of the way in which we look at the bound state in quantum mechanics, namely from the Bohr atom to the proton in the quark model. In Sec. 9.3, we introduce the quark model.

In Sec. 9.4, using the covariant harmonic wave functions, we show that the covariant harmonic oscillator wave function can explain all the peculiarities of Feynman's parton picture. We thus show that Gell-Mann's quark model and Feynman's parton model are two limiting cases of one Lorentz-covariant entity. In Sec. 9.5, we compare our calculation based on the oscillator wave function with the experimental data.

In Sec. 9.6, it is noted that the probability distribution manifests itself in electron–proton scattering. It also shows the relativistic effects when the momentum transfer becomes large during the scattering process. We study this relativistic effect using the covariant oscillator wave function introduced in Chapter 8. Finally, in Sec. 9.7, we study this relativistic effect using momentum wave functions.

9.2 Evolution of the Hydrogen Atom

Since the time of Einstein and Bohr, there has been an evolution of the way in which we look at quantum bound states, as illustrated in Fig. 9.1. The evolution took place in the following three steps:

1. The energy levels of the hydrogen atom played the pivotal role by replacing the electron orbit of the hydrogen atom with a standing wave, leading to bound states in quantum mechanics. However, the hydrogen atom cannot play a role in the Lorentz covariant world since it cannot be accelerated to a relativistic speed.

2. In 1964, the proton became a bound state of the more fundamental constituents called *quarks* (Gell-Mann, 1964). Of course, the proton is different from the hydrogen atom, but inherits the same quantum mechanics from the hydrogen atom. Unlike the hydrogen atom, the proton can be accelerated, and its speed can become very close to that of light. Thus, it is possible to study the quantum mechanics of the hydrogen atom or bound states in the Lorentz-covariant world by studying the proton in the quark model. Figure 9.1 illustrates this transition.

3. In 1969, Feynman observed that the proton, when it moves with a velocity close to that of light, appears like a collection of partons with a wide-spread momentum distribution (Feynman, 1969a,b). Partons

Evolution of the Hydrogen Atom

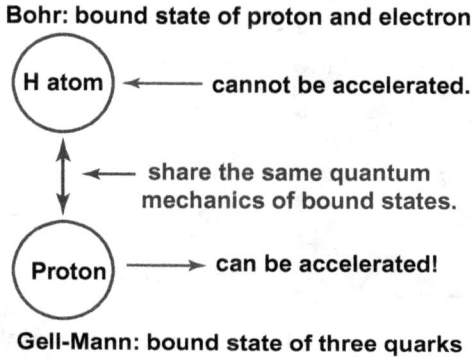

Bohr: bound state of proton and electron

(H atom) ⟵——— **cannot be accelerated.**

⟵—— **share the same quantum mechanics of bound states.**

(Proton) ——⟶ **can be accelerated!**

Gell-Mann: bound state of three quarks

Fig. 9.1. Evolution of the hydrogen atom. The electron orbit was replaced by a standing wave, but the hydrogen atom cannot be accelerated. In 1964, the proton as a bound state inherited the quantum mechanics of the hydrogen atom (Gell-Mann, 1964). The proton these days can move with a speed very close to that of light and exhibits the properties of quantum bound states in the Lorentz-covariant world.

are like free particles. Quarks and partons are the same particles but they appear differently to observers in two different reference frames. Therefore, there must be a Lorentz-covariant model for quantum bound states, as illustrated in Fig. 9.2.

At the time of Einstein and Bohr, both the proton and electron were regarded as point particles. However, the discovery of Hofstadter and McAllister changed our picture of the proton (Hofstadter and McAllister, 1955). The proton charge has an internal distribution. Within the framework of quantum electrodynamics, it is possible to calculate the Rutherford formula for electron–proton scattering when both electron and proton are point particles. Because the proton is not a point particle, there is a deviation from the Rutherford formula. We describe this deviation as the *proton form factor* which depends on the momentum transfer during electron–proton scattering.

Indeed, the study of the proton form factor has been and still is one of the central issues in high-energy physics. The proton form factor decreases as the momentum transfer increases. Its behavior is called the *dipole cut-off* meaning an inverse-square decrease, and it has been a challenging problem in quantum field theory and other theoretical models (Frazer and Fulco, 1960). Since the emergence of the quark model in 1964 (Gell-Mann, 1964),

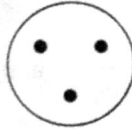

QUARKS

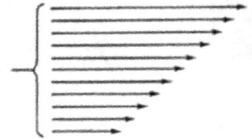

PARTONS

Two different pictures of the proton. Can they be synthesized into one Lorentz-covariant picture?

Fig. 9.2. Two distinct ways in which the proton appears in the real world. If the proton is at rest, it appears as a bound state of three quarks (Gell-Mann, 1964). If it moves with a speed close to that of light, it appears like a collection of an infinite number of partons (Feynman, 1969a,b; Bjorken and Paschos, 1969). Then the question is whether quarks and partons are two different manifestations of the same Lorentz-covariant entity. (The photo of Gell-Mann and Feynman is from the Emilio Segré Visual Archives of the American Institute of Physics.)

the hadrons are regarded as quantum bound states of quarks with space–time extensions.

Furthermore, the hadronic mass spectra indicate that the binding force between the quarks is like that of the harmonic oscillator (Feynman *et al.*, 1971). This leads us to suspect that the quark model with harmonic oscillator wave functions could explain the behavior of the proton form factor. There are indeed many papers written on this subject. We shall return to this problem in Sec. 9.6.

Another problem in high-energy physics is Feynman's parton picture (Feynman, 1969a,b; Bjorken and Paschos, 1969). If the hadron is at rest, we can approach this problem within the framework of bound-state quantum mechanics. If it moves with a speed close to that of light, it appears as a collection of an infinite number of partons, which interact with external signals incoherently. This phenomenon raises the question of whether the Lorentz boost destroys quantum coherence (Kim, 1998).

9.3 Lorentz-covariant Quark Model

In the quark model, mesons are two-body bound states of one quark and one anti-quark, and baryons are bound states of three quarks. The early successes of the quark model include the ratio of the proton–neutron electromagnetic potential and magnetic moments (Bég *et al.*, 1964). Also the hadronic mass spectra are like those of three-dimensional harmonic oscillators (Feynman *et al.*, 1971).

The question then is how the mass spectrum calculated within the framework of non-relativistic quantum mechanics is valid for this relativistic case, while ignoring the time-separation variable. For this question, the answer given in the 1971 paper of Feynman *et al.* is not satisfactory. The correct answer to this question is that Wigner's little group for massive particles is like the three-dimensional rotation group as was spelled out in Chapters 6 and 8.

Our original question is how the hydrogen atom looks to a moving observer. The question now is how much we can learn about this Bohr–Einstein issue by studying the proton in the quark model based on the three-dimensional harmonic oscillator. For the hydrogen atom, we use the Coulomb potential, while the binding force between quarks is that of the oscillator. The point is that those two different potentials share the same quantum mechanics.

For this purpose, we will need a bound-state wave function which can be Lorentz-boosted. Here, the natural choice is the harmonic oscillator wave function discussed in Chapter 8. We can start with the ground-state wave function which can be Lorentz boosted. Since the harmonic oscillator wave function is separable in the Cartesian coordinate system, we can leave out the transverse components of the wave function, and consider only the longitudinal and time-like coordinates. For this purpose, let us rewrite the wave function of Eq. (8.34) as

$$\psi_\eta(z,t) = \frac{1}{\sqrt{\pi}} \exp\left\{ -\frac{1}{4} \left[\left(e^{2\eta}(z+t)^2 + e^{-2\eta}(z-t)^2 \right) \right] \right\}, \qquad (9.1)$$

which becomes

$$\psi_0(z,t) = \frac{1}{\sqrt{\pi}} \exp\left\{ -\frac{1}{2} \left[(z^2 + t^2) \right] \right\}, \qquad (9.2)$$

for $\eta = 0$, where η is the speed parameter defined as $\tanh\eta = v/c$.

9.4 Feynman's Parton Picture

Let us go back to the two-body problem and discuss what happens to the wave function when the proton is Lorentz-boosted. For this system, we have discussed the Lorentz-squeeze problem in Chapter 8.

It is a widely accepted view that hadrons are quantum bound states of quarks having a localized probability distribution. As in all bound-state cases, this localization condition is responsible for the existence of discrete mass spectra. The most convincing evidence for this bound-state picture is the hadronic mass spectra (Feynman *et al.*, 1971; Kim and Noz, 1986).

However, this picture of bound states is applicable only to observers in the Lorentz frame in which the hadron is at rest. How would the hadrons appear to observers in other Lorentz frames? In 1969, Feynman observed that a fast-moving hadron can be regarded as a collection of many *partons* whose properties appear to be quite different from those of the quarks (Feynman, 1969a,b; Bjorken and Paschos, 1969; Başkal *et al.*, 2016). For example, the number of quarks inside a static proton is three, while the number of partons in a rapidly moving proton appears to be infinite. The question then is how the proton looking like a bound state of quarks to one observer can appear so differently to an observer in a different Lorentz frame? Feynman made the following systematic observations:

1. The picture is valid only for hadrons moving with velocity close to that of light.
2. The interaction time between the quarks becomes dilated, and partons behave as free independent particles.
3. The momentum distribution of partons becomes widespread as the hadron moves fast.
4. The number of partons seems to be infinite or much larger than that of quarks.

Because the hadron is believed to be a bound state of two or three quarks, each of the above phenomena appears as a paradox, particularly Items 2 and 3 together. How can a free particle have a wide-spread momentum distribution?

In order to resolve this paradox, let us construct the momentum–energy wave function corresponding to the Gaussian form of Eq. (9.1). The momentum–energy wave function should also be a Gaussian function. If the quarks have the four-momenta p_a and p_b, we can construct two independent four-momentum variables (Feynman *et al.*, 1971)

$$P = p_a + p_b, \qquad q = \sqrt{2}(p_a - p_b). \tag{9.3}$$

The four-momentum P is the total four-momentum and is thus the hadronic four-momentum while q measures the four-momentum separation between the quarks.

The resulting momentum-energy wave function is

$$\phi_\eta(q_z, q_0) = \left(\frac{1}{\pi}\right)^{1/2} \exp\left\{-\frac{1}{4}[e^{-2\eta}(q_z + q_0)^2 + e^{2\eta}(q_z - q_0)^2]\right\}. \quad (9.4)$$

For large values of η, we can let $q_0 = q_z$, and the wave function becomes

$$\phi_\eta(q_z) = \left(\frac{1}{\pi}\right)^{1/4} \exp\left\{-\left[e^{-2\eta}(q_z)^2\right]\right\}. \quad (9.5)$$

Because we are using here the harmonic oscillator, the mathematical form of the above momentum–energy wave function is identical to that of the space–time wave function of Eq. (9.1). The Lorentz squeeze properties of these wave functions are also the same. This aspect of the squeeze has been exhaustively discussed in the literature (Kim and Noz, 1977; Kim, 1989; Başkal *et al.*, 2015), and it is illustrated again in Fig. 9.3. The hadronic structure function calculated from this formalism is in reasonable agreement with the experimental data (Hussar, 1981) as shown in Fig. 9.4.

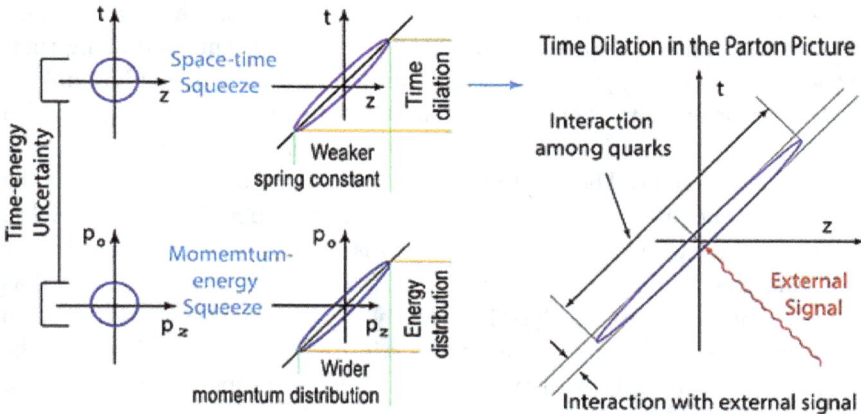

Fig. 9.3. Lorentz-squeezed space–time and momentum-energy wave functions. As the hadron's speed approaches that of light, both wave functions become concentrated along their respective positive light-cone axes. These light-cone concentrations lead to Feynman's parton picture (Kim and Noz, 1977; Kim, 1989). The external signal, since it is moving in the direction opposite to the direction of the hadron, travels along the negative light-cone axis. Thus, the interaction time of this signal with the bound state is much shorter than the period of oscillation of the quarks inside the hadron. This effect is called Feynman's time dilation (Feynman, 1969a,b; Bjorken and Paschos, 1969; Kim and Noz, 2005).

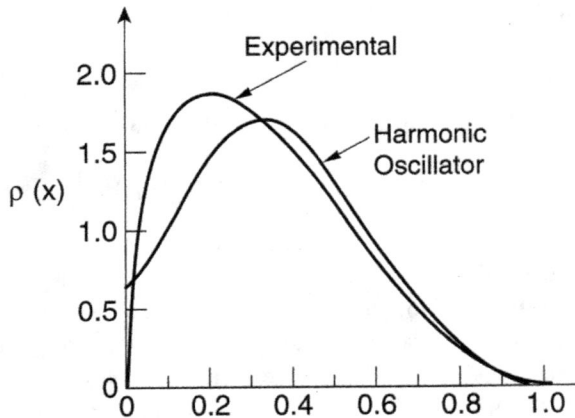

Fig. 9.4.   Parton distribution function compared with experimental data. The boosted oscillator has its peak at $x = 1/3$. This Gaussian form gives a reasonable agreement with experimental data for large values of x, but the disagreement is substantial for small values of x. This figure is from Hussar's paper (Hussar, 1981).

When the hadron is at rest with $\eta = 0$, both wave functions behave like those for the static bound state of quarks. As η increases, the wave functions become continuously squeezed until they become concentrated along their respective positive light-cone axes, as shown in Fig. 9.3. Let us look at the z-axis projection of the space–time wave function. Indeed, the width of the quark distribution increases as the hadronic speed approaches that of the speed of light. The position of each quark appears widespread to the observer in the laboratory frame, and the quarks appear like free particles.

The momentum–energy wave function is just like the space–time wave function. The longitudinal momentum distribution becomes widespread as the hadronic speed approaches the velocity of light. This is in contradiction with our expectation from non-relativistic quantum mechanics that the width of the momentum distribution is inversely proportional to that of the position wave function. Our expectation is that if quarks are free, they must have a sharply defined momenta, not a wide-spread distribution.

However, according to our Lorentz-squeezed space–time and momentum-energy wave functions, the space–time width and the momentum-energy width increase in the same direction as the hadron is boosted. This is of course an effect of Lorentz covariance. This indeed leads to the resolution of one of the quark–parton puzzles (Kim and Noz, 1977; Kim, 1989; Başkal et al., 2016).

Another puzzling problem in the parton picture is that partons appear as incoherent particles, while quarks are coherent when the hadron is at rest. Does this mean that the coherence is destroyed by the Lorentz boost (Kim, 1998)? The answer is NO, and here is the resolution to this puzzle.

When the hadron is boosted, the hadronic matter becomes squeezed and becomes concentrated in the elliptic region along the positive light-cone axis. The length of the major axis becomes expanded by e^{η}, and the minor axis is contracted by $e^{-\eta}$, as shown in Fig. 9.3 and also in Fig 8.4.

This means that the interaction time of the quarks among themselves becomes dilated. Because the wave function becomes widespread, the distance between one end of the oscillator well and the other end increases. This effect, first noted by Feynman, is universally observed in high-energy hadronic experiments (Feynman, 1969a,b; Bjorken and Paschos, 1969). The period of oscillation increases like e^{η}. On the other hand, the external signal, since it is moving in the direction opposite to the direction of the hadron, travels along the negative light-cone axis.

If the hadron contracts along the negative light-cone axis, the interaction time decreases by $e^{-\eta}$. The ratio of the interaction time to the oscillator period becomes $e^{-2\eta}$. The energy of each proton coming out of the LHC accelerator is 13 TeV. This leads to the ratio 1.25×10^{-9}. This is indeed a small number. The external signal is not able to sense the interaction of the quarks among themselves inside the hadron.

Indeed, the covariant harmonic oscillator formalism provides one Lorentz-covariant entity which produces the quark and parton models as two limiting cases as is indicated in Table 9.1.

Table 9.1 Further contents of Einstein's $E = mc^2$. The fourth row is added to Table 6.1. Indeed, the unified picture of the quark and parton models can be viewed as a further content of Einstein's energy–momentum relation in Chapter 1.

	Massive slow	Lorentz covariance	Fast massless
Energy–momentum	$mc^2 + p^2/2m$	Einstein's $E = \sqrt{(mc^2)^2 + (cp)^2}$	$E = cp$
Helicity Spin, Gauge	S_3 S_1, S_2	Wigner's little group	Helicity gauge trans.
Quarks in proton	Quark model	Covariant oscillator	Parton picture

9.5 Proton Structure Function

The quark distribution in momentum–energy space can be measured from the inelastic electron–proton scattering with one-photon exchange (Bjorken and Paschos, 1969). The measured distribution is called the proton structure function. We are now interested in how close the Gaussian form of Eq. (9.4) is to the experimental world.

First of all, in the large-η limit, the proton wave function is within the narrow elliptic region where $q_z = q_0$, and we are left with the wave function depending on only one variable. Thus, this one-variable wave function takes the form

$$\phi_\eta(q_z) = \left(\frac{1}{\pi}\right)^{1/4} \exp\left\{-\left[e^{-2\eta}(q_z)^2\right]\right\}. \tag{9.6}$$

According to Eq. (9.3),

$$p_{az} = \left(\frac{P_z}{2} + \frac{q_z}{2\sqrt{2}}\right), \qquad p_{bz} = \left(\frac{P_z}{2} - \frac{q_z}{2\sqrt{2}}\right). \tag{9.7}$$

If we introduce the parameter

$$x = \frac{p_{az}}{P_z}, \tag{9.8}$$

this is the ratio of the quark momentum to the hadronic momentum. Indeed, this variable is used for measuring the parton distribution in high-energy laboratories.

It is then possible to write the Gaussian form of Eq. (9.6) in terms of this x variable, and the quark distribution can be written as

$$\rho(x) = \exp\left[-\gamma\left(x - \frac{1}{2}\right)^2\right], \tag{9.9}$$

where the constant γ is to be determined from the level separation from the hadronic mass spectra (Feynman *et al.*, 1971). The variable x ranges from its minimum value of zero to the maximum value 1. This Gaussian form peaks at $x = 1/2$.

Before attempting to make a real contact with the experimental world, we have to face the fact that the proton is a bound state of three quarks. Within the harmonic oscillator regime, the three-body bound system can be separated into a regime of two independent oscillators. This problem was worked out in detail in the 1971 paper of Feynman *et al.* (1971). Let us reproduce their calculation.

Let x_a, x_b, x_c represent the space–time coordinates for those quarks. If there is an oscillator force between two quarks, we are led to the quadratic form

$$\left[(x_a - x_b)^2 + (x_b - x_c)^2 + (x_c - x_a)^2 \right]. \tag{9.10}$$

In order to deal with this expression, Feynman *et al.* (1971) introduced the following three variables:

$$X = \frac{x_a + x_b + x_c}{3},$$

$$r = \frac{x_a + x_b - 2x_c}{6},$$

$$s = \frac{x_b - x_a}{2}, \tag{9.11}$$

and

$$x_a = X - 2r,$$

$$x_b = X + r - \sqrt{3}s,$$

$$x_c = X + r + \sqrt{3}s. \tag{9.12}$$

In terms of the r and s variables, the quadratic form becomes

$$18 \left(r^2 + s^2 \right), \tag{9.13}$$

and does not depend on the X variable, which specifies the space–time coordinate of the proton.

As for the momentum-energy four-vectors, let us call them p_a, p_b, and p_c for the quarks a, b and c respectively, and introduce the following variables. For the momentum-energy four-vectors, we can introduce the following three variables.

$$P = p_a + p_b + p_c,$$

$$q = p_a + p_b - 2p_c,$$

$$k = \sqrt{3} \left(p_b - p_a \right). \tag{9.14}$$

Then

$$p_a = \frac{1}{3}P + \frac{1}{6}q - \frac{1}{2\sqrt{3}}k,$$

$$p_b = \frac{1}{3}P + \frac{1}{6}q + \frac{1}{2\sqrt{3}}k,$$

$$p_c = \frac{1}{3}P - \frac{1}{3}q. \tag{9.15}$$

In terms of these variables, we are led to consider the quadratic form of

$$18(q^2 + k^2). \tag{9.16}$$

This form does not depend on the variable P, which measures the momentum and energy of the proton.

If the external signal interacts with quark c, its momentum depends only on the q variable, which can be written as

$$q = P - 3p_c. \tag{9.17}$$

We can then define the x variable as

$$x = \frac{p_{cz}}{P_z}. \tag{9.18}$$

Then the quark distribution should take the form

$$\rho(x) = \left(\frac{1}{\pi\gamma}\right)^{1/2} \exp\left[-\gamma\left(x - \frac{1}{3}\right)^2\right]. \tag{9.19}$$

The constant γ is to be determined from the hadronic mass spectra based on the oscillator model (Feynman *et al.*, 1971).

This distribution is measurable and appears as the proton structure function in the real world. If the quarks interact with the external photon as point particles, the Gaussian distribution of Eq. (9.19) does not agree well with the experimental curve. On the other hand, it is known that there are various models describing the clouds surrounding the quarks.

Among the various models for this cloud effect, the valon model (Hwa, 1980; Hwa and Zahir, 1981) is one of the simplest. In 1981 (Hussar, 1981), Hussar took into account this valon model, and compared the two curves representing the experimental and boosted harmonic oscillator data. They are in reasonable agreement as shown in Fig. 9.4.

This graph may not be as accurate as we desire. However, the remarkable feature is that the Gaussian form was calculated from the proton at rest. So is the constant γ. It came from the level spacing in the hadronic mass spectra. It is remarkable that these two features manifest themselves for the proton whose speed is very close to that of the light.

There are many other models to deal with the problem of providing corrections to the parton distribution. QCD (quantum chromodynamics) is a case in point (Buras, 1980). QCD can provide corrections to the distribution, but it does not produce the distribution from which to start. The covariant harmonic oscillator function provides this starting point.

It is like the case of quantum electrodynamics. QED was quite successful in producing the Lamb shift in the hydrogen energy spectrum, but QED cannot produce the Rydberg energy levels to which the correction is made. The hydrogen energy levels are still obtained from the Schrödinger or Dirac equation with the localization condition on wave functions.

9.6 Proton Form Factor and Lorentz Coherence

Let us now consider the elastic scattering of proton and electron with one photon exchange. If the proton is a point particle, the scattering cross section can be calculated from the one-photon exchange Feynman diagram. The calculation is straight-forward if the proton is a point particle. This process is called the Rutherford scattering, and the cross section becomes the same as the classical Coulomb scattering if the proton's recoil is negligible.

As the momentum transfer becomes substantial as indicated in Fig. 9.5, the cross section deviates from that of the Rutherford scattering, as was observed first by Hofstadter and McAllister (1955). Subsequently, it was observed that the cross section decreases as

$$\frac{1}{(\text{momentum transfer})^8}. \tag{9.20}$$

This deviation comes from the fact that the proton is not a point particle and that the electric charge inside the proton is distributed with a finite radius. The portion of the scattering amplitude describing this distribution is called the proton form factor. The proton form factor should therefore decrease as

$$\frac{1}{(\text{momentum transfer})^4}. \tag{9.21}$$

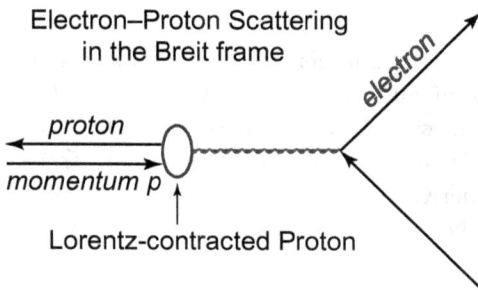

Fig. 9.5.   Breit frame. The incoming and outgoing protons move with equal magnitude of momentum in opposite directions (Kim and Noz, 1986).

This behavior of decrease is known as the dipole cut-off in the literature. This dipole cut-off and possible deviations from it constitute one of the major branches of high-energy physics. There have been in the past some far-reaching theoretical models to deal with this problem (Frazer and Fulco, 1960).

In this section, we are interested in approaching this problem using the harmonic oscillator formalism developed in Chapter 8. We shall show that the dipole cut-off is a consequence of the coherence between the contraction of the proton wave function and the decrease in the wavelength of the incoming signal.

While the formalism of Chapter 8 is largely based on the papers written by Dirac and Wigner, it is interesting to note that the same harmonic oscillator functions can be derived from those authors who attempted to understand the proton form factor. These authors were not aware of the work of Dirac and Wigner. Let us briefly review what they did.

In 1953, Yukawa was interested in constructing harmonic oscillator wave functions that can be Lorentz-transformed (Yukawa, 1953). His primary interest was in the mass spectrum produced by his Lorentz-invariant differential equation. However, at that time, his mass spectrum did not appear to have anything to do with the physical world.

After witnessing a non-zero charge radius of the proton observed by (Hofstadter and McAllister, 1955), Markov in 1956 considered using Yukawa's oscillator formalism for calculating the proton form factor (Markov, 1956).

However, the constituent particles of the oscillator wave functions were not defined at that time. Shortly after the emergence of the quark model in 1964 (Gell-Mann, 1964), Ginzburg and Man'ko considered relativistic harmonic oscillators for bound-states of quarks (Ginzburg and Man'ko, 1965).

Even though they did not mention Yukawa's 1953 paper (Yukawa, 1953), Fujimura, Kobayashi, and Namiki used the quark model based on Yukawa's relativistic oscillator wave function, to calculate the proton form factor, and obtained the dipole cut-off (Fujimura *et al.*, 1970). In the same year, Licht and Pagnamenta derived the same result using Lorentz-contracted oscillator wave functions. They used the Breit coordinate system in order to bypass the time-separation variable appearing in the covariant formalism (Licht and Pagnamenta, 1970; Kim and Noz, 1973).

In 1971, Feynman, Kislinger, and Ravndal noted that the observed hadronic mass spectra can be explained in terms of the degeneracies of three-dimensional harmonic oscillators (Feynman *et al.*, 1971), confirming the earlier suggestion made by Yukawa in 1953 (Yukawa, 1953). They quoted the paper by (Fujimura *et al.*, 1970), but they did not mention Yukawa's 1953 paper (Yukawa, 1953). This is the reason why Feynman *et al.* could not write down normalizable wave functions.

Let us go back to the formalism developed in Chapter 8. When considering the scattering of one electron and one proton by exchanging one photon, it is possible to choose the Lorentz frame in which the incoming and outgoing protons are moving in opposite directions with the same speed. Let us assume that the proton is moving along the z direction as indicated in Fig. 9.5, and let p be the magnitude of the momentum. Then the initial and final momentum–energy four-vectors are

$$(p, E) \quad \text{and} \quad (-p, E), \tag{9.22}$$

respectively, where $E = \sqrt{1 + p^2}$. The momentum transfer in this Breit frame is

$$(p, E) - (-p, E) = (2p, 0), \tag{9.23}$$

with zero energy component.

The proton form factor then becomes

$$F(p) = \int e^{2ipz} \left(\psi_\eta(z, t) \right)^* \psi_{-\eta}(z, t) \, dz \, dt. \tag{9.24}$$

If we use the ground-state harmonic oscillator wave function, this integral becomes

$$\frac{1}{\pi} \int e^{2ipz} \exp \left\{ -\cosh(2\eta) \left(z^2 + t^2 \right) \right\} \, dz \, dt. \tag{9.25}$$

The physics of $\cosh(2\eta)$ in this expression was explained in Eq. (8.54).

In the Fourier integral of Eq. (9.25), the exponential function does not depend on the t variable. Thus, after the t integration, Eq. (9.25) becomes

$$F(p) = \frac{1}{\sqrt{\pi \cosh(2\eta)}} \int e^{2ipz} \exp \left\{ -z^2 \cosh(2\eta) \right\} \, dz. \tag{9.26}$$

If we complete this integral, the proton form factor becomes

$$F(p) = \frac{1}{\cosh(2\eta)} \exp \left\{ \frac{-p^2}{\cosh(2\eta)} \right\}. \tag{9.27}$$

If we use the expression of $\cosh(2\eta)$ given in Eq. (8.54), this proton form factor becomes

$$F(p) = \frac{1}{1 + 2p^2} \exp\left(\frac{-p^2}{1 + 2p^2}\right), \qquad (9.28)$$

which decreases as $1/p^2$ for large values of p.

In order to illustrate the effect of the role of this Lorentz contraction in more detail, let us perform the integral of Eq. (9.26) without the contraction factor $\cosh(2\eta)$. This means that the wave function $\psi_\eta(z, t)$ in the Eq. (9.24) is replaced by the Gaussian form $\psi_0(z, t)$ of Eq. (9.1). With this non–squeezed wave function, the Fourier integral becomes

$$G(p) = \int e^{2ipz} \left(\psi_0(z, t)\right)^* \psi_0(z, t) \, dz \, dt. \qquad (9.29)$$

The result of this integral is

$$G(p) = \frac{1}{\sqrt{\pi}} \exp\left(-p^2\right). \qquad (9.30)$$

This leads to a Gaussian cutoff of the proton form factor. This does not happen in the real world, and the calculation of $G(p)$ is for an illustrative purpose only.

Let us go back to the Fourier integrals of Eqs. (9.24) and (9.29). The only difference is the $\cosh(2\eta)$ factor in Eq. (9.24). This factor is in the normalization constant and comes from the integration over the t variable which does not affect the Fourier integral.

However, it causes the Gaussian width to shrink by $1/\sqrt{2}p$ for large values of p. The wavelength of the sinusoidal factor is inversely proportional to the momentum $2p$. Thus, both the Gaussian width and the wavelength of the incoming signal shrink at the same rate of $1/p$ as p becomes large. Without this coherence, the cutoff is Gaussian as noted in Eq. (9.30). The effect of this Lorentz coherence is illustrated in Fig. 9.6.

There still is a gap between $F(p)$ of Eq. (9.28) and the real world. Before comparing this expression with the experimental data, we have to realize that there are three quarks inside the proton with two oscillator modes.

One of the modes goes through the Lorentz coherence process discussed in this section. The other mode goes through the contraction process given in Eq. (8.54). The net effect is

$$F_3(p) = \left(\frac{1}{1 + 2p^2}\right)^2 \exp\left(\frac{-p^2}{1 + 2p^2}\right). \qquad (9.31)$$

This will lead to the desired dipole cut-off of $(1/p^2)^2$.

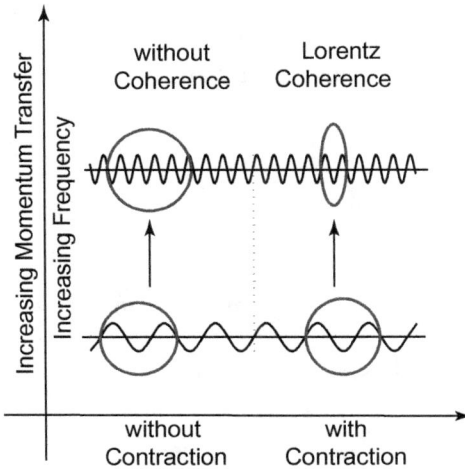

Fig. 9.6. Coherence between the wavelength and the proton size. Referring back to Fig. 9.5, the proton sees the incoming photon. The wavelength of this photon becomes smaller for increasing momentum transfer. If the proton size remains unchanged, there is a rapid oscillation cutoff in the Fourier integral for the form factor leading to a Gaussian cutoff. However, if the proton size decreases coherently with the wavelength, there are no oscillation effects, leading to a polynomial decrease of the form factor (Kim and Noz, 1986; Başkal *et al.*, 2015).

In addition, the effect of the quark spin should be addressed. There are also reports of deviations from the exact dipole cut-off. There have been attempts to study the proton form factors based on the four-dimensional rotation group with an imaginary time coordinate. There are also many papers based on the lattice QCD. A brief list of the references to these efforts is given by (Kim and Noz, 2011).

The purpose of this section was limited to studying in detail the role of Lorentz coherence in keeping the proton form factor from the steep Gaussian cutoff in the momentum transfer variable. The coherence problem is one of the primary issues of the current trend in physics.

9.7 Coherence in Momentum-Energy Space

We are now interested in how Lorentz coherence manifests itself in momentum–energy space. We start with the Lorentz-squeezed wave function in momentum–energy space, which can be written as

$$\phi_\eta \left(q_z, q_0\right) = \frac{1}{2\pi} \int e^{-i(q_z z - q_0 t)} \psi_\eta(z, t) dt \, dz. \tag{9.32}$$

This is a Fourier transformation of the Lorentz-squeezed wave function of Eq. (9.1), where q_z and q_0 are Fourier conjugate variables to z and t respectively. The result of this integral is

$$\phi_\eta(q_z, q_0) = \frac{1}{\sqrt{\pi}} \exp\left\{ -\frac{1}{4}\left[e^{-2\eta}(q_z + q_0)^2 + e^{2\eta}(q_z - q_0)^2 \right] \right\}. \quad (9.33)$$

In terms of this momentum–energy wave function, the proton form factor of Eq. (9.24) can be written as

$$F(p) = \int \phi^*_{-\eta}(q_0, q_z - p)\,\phi_\eta(q_0, q_z + p)\,dq_0\,dq_z. \quad (9.34)$$

The evaluation of this integral leads to the proton form factor $F(p)$ given in Eq. (9.28).

In order to see the effect of the Lorentz coherence, let us look at the two wave functions in Fig. 9.7. The integral is carried out over the $q_z\,q_0$ plane. As the momentum p increases, the two wave functions become separated. Without the Lorentz squeeze, the wave functions do not overlap, and this leads to a sharp Gaussian cutoff as in the case of $G(p)$ of Eq. (9.30).

On the other hand, the squeezed wave functions have an overlap as shown in Fig. 9.7, and this overlap becomes smaller as p increases. This leads to a slower polynomial cutoff (Kim and Noz, 1986, 2011).

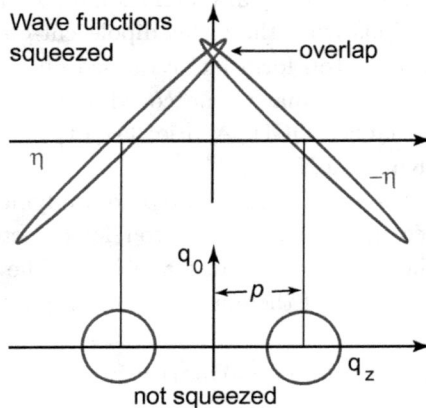

Fig. 9.7. Lorentz coherence in the momentum-energy space. Both squeezed and non-squeezed wave functions are given. As p increases, the two wave functions in Eq. (9.34) become separated. Without the squeeze, there are no overlaps. This leads to a Gaussian cutoff. The squeezed wave functions maintain an overlap, leading to a slower polynomial cutoff (Kim and Noz, 1986).

The discovery of the non-zero size of the proton opened a new era of physics (Hofstadter and McAllister, 1955; Hofstadter, 1956). The proton is no longer a point particle. One way to measure its internal structure is to study the proton–electron scattering amplitude with one photon exchange, and its dependence on the momentum transfer. The deviation from the case with the point-particle proton is called the proton form factor.

On the experimental front, the dipole cut-off has been firmly established. Yes, there are also experimental results indicating deviations from this dipole behavior (Alkofer *et al.*, 2005; Matevosyan *et al.*, 2005). However, in the present section, no attempts have been made to review all the papers written on the corrections. From the theoretical point of view, those deviations are corrections from the basic dipole behavior.

While the study of the proton form factor is still a major subject in physics, it is gratifying to note that the proton's dipole cut-off comes from the coherence between the Lorentz contraction of the proton's longitudinal size and the decrease in the wavelength of the incoming signal.

Chapter 10

Feynman's Rest of the Universe

In Chapter 8, we constructed a set of harmonic oscillator wave functions that can be Lorentz-boosted. These wave functions carry the quantum probability interpretation, after the integration over the t variable. This variable is the time-separation variable which is thoroughly hidden in the present form of quantum mechanics.

Dirac's Gaussian form of 1945 (Dirac, 1945b) has this variable, but he did not give it a physical interpretation. In their paper of 1971, (Feynman *et al.*, 1971) pointed out clearly that this time separation variable is applicable between the constituent particles inside the hadron. However, they chose to ignore this variable because they did not know how to deal with it. This is not what we expect from Feynman's paper. Yet, he still deserves our respect for pointing out that the problem exists with this variable.

The purpose of this chapter is to provide the resolution to this problem within the existing framework of quantum mechanics and special relativity. The problem is with the transition from the wave function to the measurable probability distribution. This transition process is called the density matrix in non-relativistic quantum mechanics. The question is whether this process is possible in the Lorentz-covariant world, particularly within the framework of the covariant harmonic oscillator formalism.

In his book on statistical mechanics (Feynman, 1998), whose first edition was published in 1972, Feynman makes the following statement about the density matrix.

147

When we solve a quantum-mechanical problem, what we really do is divide the universe into two parts — the system in which we are interested and the rest of the universe. We then usually act as if the system in which we are interested comprised the entire universe. To motivate the use of density matrices, let us see what happens when we include the part of the universe outside the system.

Here Feynman summarized the content of von Neumann's book entitled *Die mathematische Grundlagen der Quanten-mechanik* published in 1932 (von Neumann, 1932), and translated into English as *Mathematical Foundation of Quantum Mechanics* in 1955 (Von Neumann, 1996).

In this chapter, we study Feynman's rest of the universe using two coupled harmonic oscillators whose quantum mechanics and density matrix are well understood (Feynman, 1998; Han *et al.*, 1999). It will be noted that the wave function for this system is identical with the covariant oscillator wave function discussed in Chapter 8.

The issue is how to transform wave functions to measurable probability density, when not all the variables are measured. In the oscillator system, let x_1 and x_2 be the coordinate variables for the two-oscillator system, and let x_2 be the unmeasurable variable. This unmeasurable variable is in Feynman's rest of the universe.

Then the covariant harmonic oscillator has the same set of wave functions as the two-oscillator system discussed in Chapter 8, if the x_1 and x_2 coordinates are replaced by the z and t variables of the covariant oscillators.

The coupled oscillator provides the mathematical tool for many other branches of physics of current interest, including two-photon systems in quantum optics (Yuen, 1976; Kim and Noz, 1991; Walls and Milburn, 2008). In the two-photon system, it is possible to ignore one of the photons. This case serves as an excellent example of Feynman's rest of the universe (Yurke and Potasek, 1987; Ekert and Knight, 1989).

10.1   Introduction

Thanks to its mathematical simplicity, the harmonic oscillator provides soluble models in many branches of physics. It often gives a clear illustration of abstract ideas. In Chapter 8, we have seen that the oscillator wave functions can tell us how to boost localized waves of bound states in quantum mechanics in the Lorentz-covariant world.

In quantum mechanics, the single oscillator can be excited in the following three ways:

1. Excitation through the Schrödinger equation with discrete energy levels.
2. Coherent and squeeze preserving the minimum uncertainty product.
3. Thermal excitations. This is a statistical excitation.

Unlike the first two cases, the thermal excitation cannot be made through linear combinations of the wave functions. This excitation is a statistical ensemble requiring the concept of density matrix. The density matrix for the harmonic oscillator takes the form

$$\rho(x_1, x_2) = (1 - e^{-\hbar\omega/k_B T}) \sum_n e^{-n\hbar\omega/k_B T} \chi_n(x_1)\chi_n(x_2), \qquad (10.1)$$

where $\hbar\omega$ is the energy spacing of the oscillator, and k_B is Boltzmann's constant. T is the absolute temperature. We shall hereafter use the notation

$$\frac{\hbar\omega}{k_B T} \rightarrow \frac{1}{T}, \qquad (10.2)$$

which means the unit of the temperature is $\hbar\omega/k_B$,

$$\rho(x_1, x_2) = (1 - e^{-1/T}) \sum_n e^{-n/T} \chi_n(x_1)\chi_n(x_2). \qquad (10.3)$$

The density matrix was first introduced in 1932 by John von Neumann (von Neumann, 1932). Why do we need the density matrix? The wave function in quantum mechanics accommodates the superposition principle. However, the process of converting this function to a probability distribution appears to be taking the product of the wave function and its complex conjugate.

However, the process is more complicated. Since quantum mechanics is based on the inherent limitation of measurements, the density matrix depends heavily on how measurements are taken in laboratories. Sometimes, we measure all the variables allowed by the quantum system, and sometimes we do not. The density matrix allows us to deal with this problem. For the unmeasured variables, we sum or integrate the density matrix over those unmeasurable variables (Von Neumann, 1996; Fano, 1957; Wigner and Yanase, 1963; Feynman, 1998; Blum, 2012).

As for the unobservable variables, the time-separation variable is a case in point. It was known to both Einstein and Bohr that there is a space-like separation between the proton and the electron in the hydrogen atom. This separation is known as the Bohr radius. When the system is Lorentz-boosted, the space separation picks up a time-like component, as

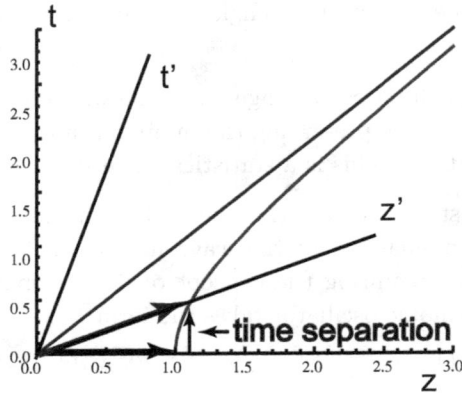

Fig. 10.1.   Time separation in relativity. In relativity, there is a time separation wherever there is a space separation. The Bohr radius is a space separation. It was (Feynman *et al.*, 1971) who pointed out the existence of this variable, but they said they did not know what to do with it in their oscillator formalism.

is illustrated in Fig. 10.1. Most certainly, this variable is in Feynman's rest of the universe.

In this chapter, we present first a model of Feynman's rest of the universe using two coupled harmonic oscillators (Han *et al.*, 1999). If one of those two oscillators is not observed, it belongs to the rest of the universe. The system of the covariant harmonic oscillator presented in Chapter 8 has two variables, namely z for the longitudinal separation and t for the time-like separation.

It will then be noted that these two physical systems share the same form for their wave functions. The physics of the coupled oscillator system can therefore be translated into the covariant harmonic oscillators. The physics of the unobservable variable in the coupled oscillator system is well understood. Thus, the same physics can be applied to the time-separation variable never measured in the present form of quantum mechanics.

Among the physical effects of the unobserved variables are an increase in entropy and a temperature rise, as well as the addition of statistical uncertainty. We should then know how to deal with the time-separation variable and its effect on the hadronic system.

In Sec. 10.2, we consider two coupled harmonic oscillators. It is possible to couple them by a canonical transformation. This coupled oscillator system shares essentially the same mathematical formalism as the covariant harmonic oscillators. It is thus possible to transfer physical interpretations from the coupled oscillator system to the covariant harmonic oscillators. This aspect is discussed in Sec. 10.3.

In Sec. 10.4, the density matrix is constructed for the two oscillator system whose quantum mechanics is well established. It is then noted that the wave function for this oscillator system is the same mathematically as that for the covariant harmonic oscillator, leading to the same form for both the coupled oscillators and the covariant harmonic oscillator. It is possible to study what happens to the coupled oscillators within the framework of the present form of quantum mechanics. Then the physics of these two oscillators can become applicable to the covariant harmonic oscillators. The physics in question is what happens if one of the oscillators is not observed or *belongs to Feynman's rest of the universe.*

In Sec. 10.5, it is noted the time-separation variable in the covariant oscillator belongs to the rest of the universe, like the unobserved oscillator in the two-oscillator system. This effect will lead to the entropy of the system. The entropy will be zero for the oscillator at rest, and will increase as the oscillator gains speed. This entropy increase can be translated into temperature rise. We study this temperature rise in Sec. 10.6. This allows us in Sec. 10.7, to explain the transition from bound states of hadrons to a plasma state at a sufficiently high temperature. This phase transition is another explanation of the transition from the quark model to the parton picture discussed in Chapter 9.

As for the uncertainty relation, it is more convenient to use the Wigner function which can be derivable from the density matrix. In Sec. 10.8, we derive the Wigner function from the density matrix given in Sec. 10.4. This Wigner function will tell us how much is from the quantum uncertainty and how much is from the rest of the universe. It is shown in Sec. 10.9 how the fundamental uncertainty remains Lorentz invariant.

10.2 Coupled Oscillators and Covariant Oscillators

Let us start with two uncoupled oscillators. The Hamiltonian for this system is

$$H = \frac{1}{2}\left(p_1^2 + x_1^2\right) + \frac{1}{2}\left(p_2^2 + x_2^2\right),\qquad(10.4)$$

where

$$p_1 = -i\frac{\partial}{\partial x_1},\qquad p_2 = -i\frac{\partial}{\partial x_2}.$$

If we introduce the variables

$$x_+ = \frac{x_1 + x_2}{\sqrt{2}}, \qquad x_- = \frac{x_1 - x_2}{\sqrt{2}},$$

$$p_+ = \frac{p_1 + p_2}{\sqrt{2}}, \qquad p_- = \frac{p_1 - p_2}{\sqrt{2}}. \tag{10.5}$$

The Hamiltonian becomes

$$H = \frac{1}{2} \left[p_+^2 + x_+^2 \right] + \frac{1}{2} \left[p_-^2 + x_-^2 \right]. \tag{10.6}$$

The resulting ground-state wave function is

$$\psi(x_1, x_2) = \frac{1}{\sqrt{\pi}} \exp \left\{ -\frac{1}{2} \left(x_+^2 + x_-^2 \right) \right\}. \tag{10.7}$$

This aspect of the two uncoupled oscillators is well known. The question is what happens when they become coupled.

The canonical way to couple these two oscillators is to apply the coordinate transformation

$$\begin{pmatrix} x_+ \\ p_- \end{pmatrix} \rightarrow e^{-\eta} \begin{pmatrix} x_+ \\ p_- \end{pmatrix}, \qquad \begin{pmatrix} x_- \\ p_+ \end{pmatrix} \rightarrow e^{\eta} \begin{pmatrix} x_- \\ p_+ \end{pmatrix}. \tag{10.8}$$

This transformation leads to the Hamiltonian of the form

$$H_\eta = \frac{1}{2} \left[e^{2\eta} p_+^2 + e^{-2\eta} x_+^2 \right] + \frac{1}{2} \left[e^{-2\eta} p_-^2 + e^{2\eta} x_-^2 \right], \tag{10.9}$$

and the wave function of Eq. (10.7) to

$$\psi_\eta(x_1, x_2) = \frac{1}{\sqrt{\pi}} \exp \left\{ -\frac{1}{4} \left[e^{-2\eta} (x_1 + x_2)^2 + e^{2\eta} (x_1 - x_2)^2 \right] \right\}. \tag{10.10}$$

This canonical transformation is illustrated in Fig. 10.2. From this figure, it is quite safe to say that the canonical transformation in this case is a *squeeze* transformation.

While the Hamiltonians of Eqs. (10.6) and (10.9) take two different forms, we are led to the question of whether there is a quantity invariant under this canonical transformation. For this purpose, let us consider the Hamiltonian of the form

$$H_{\mathrm{inv}} = \frac{1}{2} (p_1^2 + x_1^2) - \frac{1}{2} (p_2^2 + x_2^2). \tag{10.11}$$

This is the expression for the energy of the first oscillator minus that of the second oscillator in this two-oscillator system. If we write this form in terms of the $x_\pm$ and $p_\pm$ variables,

$$H_{\mathrm{inv}} = p_+ p_- + x_+ x_-. \tag{10.12}$$

This form is invariant under the canonical transformation of Eq. (10.8).

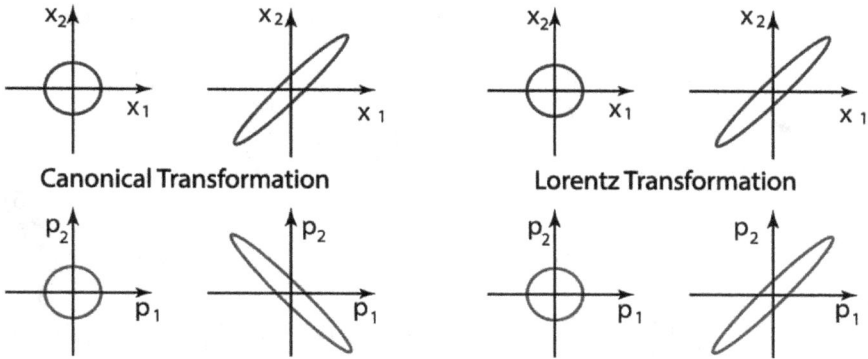

Fig. 10.2. Canonical and Lorentz transformations. In the canonical transformation, the momentum coordinates are inversely proportional to their conjugate coordinates. In the Lorentz transformation, the momentum space and momentum coordinate transform in the same way, as is seen in Fig. 9.3 of Chapter 9.

In addition, let us consider the transformation

$$\begin{pmatrix} x_+ \\ p_+ \end{pmatrix} \to e^{-\eta} \begin{pmatrix} x_+ \\ p_+ \end{pmatrix}, \qquad \begin{pmatrix} x_- \\ p_- \end{pmatrix} \to e^{\eta} \begin{pmatrix} x_- \\ p_- \end{pmatrix}. \tag{10.13}$$

This is not a canonical transformation. The space and momentum variables are transformed in the same way, as in the case of Einstein's Lorentz boost. Thus, we choose to call this transformation the Lorentz transformation, and illustrate it in Fig. 10.2. Indeed, this is how the covariant oscillator functions are Lorentz-boosted as shown in Chapter 8.

The invariant Hamiltonian was considered first by Yukawa (1953), and then by Feynman *et al.* (1971). Its Lorentz-covariant wave function enables us to construct a representation of Wigner's little group for a localized bound-state wave function as seen in Chapter 8. Thus, it enables us to explain Gell-Mann's quark model and Feynman's parton model as two special cases of one Lorentz covariant entity as shown in Chapter 9.

10.3   Entangled Oscillators

The wave function for the starting Hamiltonian of Eq. (10.4) is separable in the x_1 and x_2 variables, and can be written as

$$\chi_n (x_1) \chi_m (x_2), \tag{10.14}$$

where χ_n is the oscillator wave function in the nth excited state. In this chapter, we will be mostly interested in the case where the $\chi_m (x_2)$ is in the ground state with $m = 0$. Thus, the wave function takes the form

$$\psi_0^n(x_1, x_2) = \chi_n(x_1)\chi_0(x_2)$$

$$= \left[\frac{1}{\pi 2^n n!}\right]^{1/2} H_n(x_1) \exp\left\{-\frac{1}{4}[(x_1 + x_2)^2 + (x_1 - x_2)^2]\right\}.$$

$$(10.15)$$

For $n = 0$, this wave function becomes simplified to the Gaussian form

$$\left[\frac{1}{\pi}\right]^{1/2} \exp\left\{-\frac{1}{4}(x_1 + x_2)^2 + (x_1 - x_2)^2\right\}. \tag{10.16}$$

These wave functions are separable in the x_1 and x_2 variables.

If canonical-transformed according to Eq. (10.8), the x_1 and x_2 variables in the wave functions are replaced according to

$$x_1 \rightarrow x_1 \cosh\eta - x_2 \sinh\eta, \qquad x_2 \rightarrow x_2 \cosh\eta - x_1 \sinh\eta. \tag{10.17}$$

The wave functions are non-separable and become entangled. According to Eq. (8.45) of Chapter 8, the wave function of Eq. (10.15) becomes entangled to

$$\psi_\eta^n(x_1, x_2) = \left(\frac{1}{\cosh\eta}\right)^{(n+1)} \sum_k \left[\frac{(n+k)!}{n!k!}\right]^{1/2} (\tanh\eta)^k \chi_{n+k}(x_1) \chi_k(x_2).$$

$$(10.18)$$

If $n = 0$, this expression becomes simplified to

$$\psi_\eta^0(x_1, x_2) = \left(\frac{1}{\cosh\eta}\right)^{1/2} \sum_k (\tanh\eta)^k \chi_k(x_1) \chi_k(x_2). \tag{10.19}$$

This expression is still entangled, and can also be written in the Gaussian form given in Eq. (10.10).

Indeed, the canonical-transformed wave function of x_1 and x_2 becomes the Lorentz-boosted wave function if these variables are replaced by z and t. The canonical transformation of two coupled oscillator *entangles* two oscillator variables, in the way the space and the time variables are entangled in the system of covariant harmonic oscillators discussed in Chapter 8. We are interested in what happens to this coupled oscillator system where the second variable x_2 is not measured. This will then tell us about the case of the covariant harmonic oscillators where the t variable belongs to Feynman's rest of the universe.

10.4 Density Matrix and Entropy

Density matrices play very important roles in quantum mechanics and statistical mechanics, especially when not all measurable variables are measured in laboratories. But they are not fully discussed in the existing textbooks. On the other hand, since there are books and review articles on this subject (Fano, 1957; Blum, 2012), it is not necessary to give here a full-fledged introduction to the theory of density matrices.

In this section, we are only interested in studying what Feynman said about the density matrix using coupled oscillators (Feynman, 1998). If all the variables are measured, the density matrix is defined as

$$\rho\left(x_1, x_2; x_1', x_2'\right) = \psi\left(x_1, x_2\right)\psi^*\left(x_1', x_2'\right), \tag{10.20}$$

and this form is called the density matrix for the pure state in the space of both x_1 and x_2.

On the other hand, if we do not make measurements in the x_2 space, we have to construct the matrix $\rho(x_1, x_1')$ by integrating over the x_2 variable:

$$\rho(x_1, x_1') = \int \rho\left(x_1, x_2; x_1', x_2\right) \, dx_2. \tag{10.21}$$

It is possible to evaluate the above integral for $n = 0$. Since we are now dealing only with the x_1 variable, we shall drop its subscript and use x for the variable for the world in which we are interested. The result of the above integration is

$$\rho(x, x') = \frac{1}{\sqrt{\pi \cosh(2\eta)}} \exp\left\{-\frac{1}{4}\left[\frac{(x+x')^2}{\cosh(2\eta)} + (x-x')^2 \cosh(2\eta)\right]\right\}. \tag{10.22}$$

With this expression, we can check the trace relations for the density matrix. Indeed the trace integral

$$Tr(\rho) = \int \rho(x, x) \, dx \tag{10.23}$$

becomes 1, as in the case of all density matrices.

As for $Tr(\rho^2)$, the result of the trace integral becomes

$$Tr(\rho^2) = \int \rho(x, x')\rho(x', x) \, dx' \, dx = \frac{1}{\cosh \eta}. \tag{10.24}$$

This is less than one. This is consistent with the general theory of density matrices. If $\eta = 0$, the oscillators become decoupled, and the first oscillator is totally independent of the second oscillator. The first oscillator is in a pure state, and $Tr(\rho^2)$ becomes 1. This result is also consistent with general theory of density matrices (Fano, 1957; Blum, 2012).

We can study the density matrix in terms of the orthonormal expansion given in Eq. (10.18). The pure-state density matrix defined in this case takes the form

$$\rho(x_1, x_2; x_1', x_2') = \sum_{k,k'} A_k(\eta) A_{k'}^*(n) \chi_{n+k}(x_1) \chi_k(x_2) \chi_{n+k'}^*(x_1') \chi_{k'}^*(x_2').$$
(10.25)

If x_2 is not measured, the density matrix becomes

$$\rho(x, x') = \sum_k |A_k(\eta)|^2) \chi_{n+k}(x) \chi_{n+k}^*(x'),$$
(10.26)

after integration over the x_2 variable. Here again, we use x for x_1.

The entropy then becomes

$$S = -\sum_k |A_k(\eta)|^2 \ln\left(|A_k(\eta)|^2\right).$$
(10.27)

We measure the entropy in units of Boltzmann's constant k. The entropy is zero for a pure state, and increases as the system becomes impure. Like $Tr(\rho^2)$, this quantity is a measure of our ignorance about the rest of the universe.

10.5 Entropy and Lorentz Transformation

We can now go back to the space–time wave function of Chapter 8. If the z and t variables are both measurable, we can construct the density matrix

$$\rho_\eta^n(z, t; z', t') = \psi_\eta^n(z, t) \left(\psi_\eta^n(z', t')\right)^*,$$
(10.28)

with the covariant harmonic oscillators given in Chapter 8, which takes the form

$$\psi_\eta^n(z, t) = \left(\frac{1}{\cosh\eta}\right)^{(n+1)} \sum_k \left[\frac{(n+k)!}{n!k!}\right]^{1/2} (\tanh\eta)^k \chi_{n+k}(z) \chi_k(t).$$
(10.29)

Since the oscillator wave functions are all real, we shall hereafter drop the $*$ sign for complex conjugate.

This form satisfies the pure-state condition $\rho^2 = \rho$ which can be written explicitly as

$$\rho_\eta^n(z, t; z', t') = \int \rho_\eta^n(z, t; r, s) \rho_\eta^n(r, s; z', t') \, dr \, ds.$$
(10.30)

However, there are at present no measurement theories which accommodate the time-separation variable t. Thus, we can take the trace of the ρ

matrix with respect to the t variable. Then the resulting density matrix is

$$\rho_\eta^n(z, z') = \int \psi_\eta^n(z, t)\psi_\eta^n(z', t) \, dt$$

$$= \left(\frac{1}{\cosh\eta}\right)^{2(n+1)} \sum_k \frac{(n+k)!}{n!k!}(\tanh\eta)^{2k}\chi_{n+k}(z)\chi_{k+n}(z').$$

$$(10.31)$$

The trace of this density matrix is one, as

$$\left(\frac{1}{\cosh\eta}\right)^{2(n+1)} \sum_k \frac{(n+k)!}{n!k!}(\tanh\eta)^{2k} = 1. \qquad (10.32)$$

On the other hand, the trace of ρ^2 is less than one, as

$$Tr\left(\rho^2\right) = \int \rho_\eta^n(z, z')\rho_\eta^n(z', z) \, dz \, dz'$$

$$= \left(\frac{1}{\cosh\eta}\right)^{4(n+1)} \sum_k \left[\frac{(n+k)!}{n!k!}\right]^2 (\tanh\eta)^{4k}. \qquad (10.33)$$

This is less than one and is due to the fact that we do not know how to deal with the time-like separation in the present formulation of quantum mechanics. Our knowledge is less than complete. The time separation variable is in Feynman's rest of the universe.

The standard way to measure this ignorance is to calculate the entropy defined as (Von Neumann, 1996; Fano, 1957; Blum, 2012)

$$S = -\text{Tr}\left(\rho\ln(\rho)\right). \qquad (10.34)$$

If we pretend to know the distribution along the time-like direction and use the pure-state density matrix given in Eq. (10.28), then the entropy is zero. However, if we do not know how to deal with the distribution along t, then we should use the density matrix of Eq. (10.31) to calculate the entropy, and the result is (Kim and Wigner, 1990a)

$$S = (n+1)\left[\left(\cosh^2\eta\right)\ln\left(\cosh^2\eta\right) - (\sinh^2\eta)\ln\left(\sinh^2\eta\right)\right]$$

$$-\left(\frac{1}{\cosh\eta}\right)^{2(n+1)} \sum_k \frac{(n+k)!}{n!k!}\ln\left[\frac{(n+k)!}{n!k!}\right](\tanh\eta)^{2k}. \qquad (10.35)$$

In terms of the velocity v of the hadron

$$S = -(n+1)\left\{\ln\left[1 - \left(\frac{v}{c}\right)^2\right] + \frac{(v/c)^2\ln(v/c)^2}{1 - (v/c)^2}\right\}$$

$$-\left[1 - \left(\frac{v}{c}\right)^2\right]\sum_k \frac{(n+k)!}{n!k!}\ln\left[\frac{(n+k)!}{n!k!}\right]\left(\frac{v}{c}\right)^{2k}, \qquad (10.36)$$

where

$$\frac{v}{c} = \tanh \eta. \tag{10.37}$$

For the ground state with $n = 0$, the density matrix of Eq. (10.31) becomes

$$\rho_\eta(z, z') = \left(\frac{1}{\cosh \eta}\right)^2 \sum_k (\tanh \eta)^{2k} \chi_k(z) \chi_k^*(z'), \tag{10.38}$$

and the entropy becomes

$$S = \left(\cosh^2 \eta\right) \ln \left(\cosh^2 \eta\right) - \left(\sinh^2 \eta\right) \ln \left(\sinh^2 \eta\right). \tag{10.39}$$

Let us go back to the wave function of Eq. (10.15). As is illustrated in Fig. 10.3, its localization property is dictated by the Gaussian factor which corresponds to the ground-state wave function. For this reason, we expect that much of the behavior of the density matrix or the entropy for the nth excited state will be the same as that for the ground state with $n = 0$.

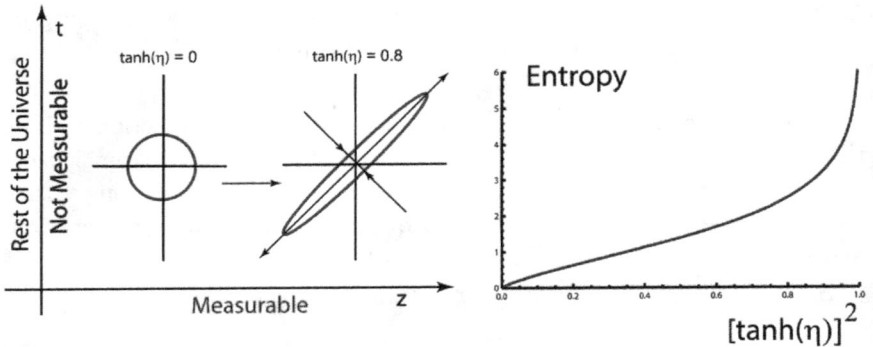

Fig. 10.3. Localization property in the zt plane. When the hadron is at rest, the Gaussian form is concentrated within a circular region specified by $(z+t)^2 + (z-t)^2 = 1$. As the hadron gains speed, the region becomes deformed to $e^{-2\eta}(z+t)^2 + e^{2\eta}(z-t)^2 = 1$. Since it is not possible to make measurements along the t direction, we have to deal with information that is less than complete. This does not cause problems when the hadron is at rest, because the t dependence in the density matrix is separable and can be integrated out. When the t separation is not measured as in the case of the Schrödinger quantum mechanics, the entropy of the system becomes non-zero and becomes increased as the hadron gains speed.

For the ground state, the wave function becomes

$$\psi_\eta(z,t) = \left[\frac{1}{\pi}\right]^{1/2} \exp\left\{-\frac{1}{4}\left(e^{-2\eta}(z+t)^2 + e^{2\eta}(z-t)^2\right)\right\}. \qquad (10.40)$$

The density matrix is

$$\rho(z,z') = \left(\frac{1}{\pi\cosh(2\eta)}\right)^{1/2} \exp\left\{-\frac{1}{4}\left[\frac{(z+z')^2}{\cosh(2\eta)} + (z-z')^2\cosh(2\eta)\right]\right\}. \qquad (10.41)$$

The probability distribution $\rho(z,z)$ becomes

$$\rho(z,z) = \left(\frac{1}{\pi\cosh(2\eta)}\right)^{1/2} \exp\left(\frac{-z^2}{\cosh(2\eta)}\right). \qquad (10.42)$$

The width of the distribution becomes $\sqrt{\cosh(2\eta)}$, and becomes widespread as the hadronic speed increases. Likewise, the momentum distribution becomes widespread as can be seen in Fig. 10.2. This simultaneous increase in the momentum and position distribution widths is seen in the parton phenomenon in high-energy physics (Feynman, 1969a,b; Kim and Noz, 1977, 1986).

10.6   Hadronic Temperature

If the single oscillator becomes thermally excited, the density matrix becomes

$$\rho_T(z,z') = \left(1 - e^{-1/T}\right) \sum_k e^{-1/T}\phi_k(z)\phi_k^*(z'). \qquad (10.43)$$

If we compare this expression with the density matrix of Eq. (10.38), we are led to

$$\tanh^2\eta = \exp\left(-1/T\right), \qquad (10.44)$$

and to

$$T = \frac{-1}{\ln\left(\tanh^2\eta\right)}. \qquad (10.45)$$

The temperature can be calculated as a function of $\tanh(\eta)$, and this calculation is plotted in Fig. 10.4.

Earlier in Eq. (10.37), we noted that $\tanh(\eta)$ is v/c. Thus, the oscillator becomes thermally excited as it moves, as is illustrated in Fig. 10.4.

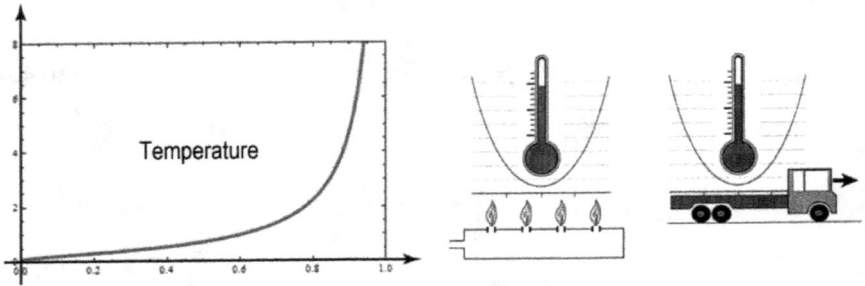

Fig. 10.4.   Boiling quarks become partons. When the hadron gains speed, the temperature of the system rises according to Eq. (10.45). The quarks will boil, and they will go through a phase transition to partons as is indicated in Fig. 10.5.

Let us look at the velocity dependence of the temperature again. It is almost proportional to the velocity from $\tanh(\eta) = 0$ to 0.7, and again from $\tanh(\eta) = 0.9$ to 1 with different slopes. We shall return to this issue in Sec. 10.7 in connection with the transition from the quark model to the parton model which was discussed in Chapter 8.

While the physical motivation for this chapter is based on Feynman's time separation variable (Feynman *et al.*, 1971) and his rest of the universe (Feynman, 1998), we should note that many authors discussed field theoretic approach to derive the density matrix of Eq. (10.38). Among them are two-mode squeezed states of light (Yuen, 1976; Yurke *et al.*, 1986; Kim and Noz, 1991; Han *et al.*, 1993) and thermo-field-dynamics (Fetter and Walecka, 2003; Umezawa *et al.*, 1982; Oz-Vogt *et al.*, 1991).

The mathematics of two-mode squeezed states is the same as that for the covariant harmonic oscillator formalism discussed in this chapter (Dirac, 1963; Yurke *et al.*, 1986; Kim and Noz, 1991; Han *et al.*, 1993). Instead of the z and t coordinates, there are two measurable photons. If we choose not to observe one of them, it belongs to Feynman's rest of the universe (Han *et al.*, 1999; Yurke and Potasek, 1987; Ekert and Knight, 1989).

Another remarkable feature of two-mode squeezed states of light is that its formalism is identical to that of thermo-field-dynamics (Fetter and Walecka, 2003; Ojima, 1981; Umezawa *et al.*, 1982; Oz-Vogt *et al.*, 1991). The temperature is in this case related to the squeeze parameter. It is therefore possible to define the temperature of a Lorentz-squeezed hadron within the framework of the covariant harmonic oscillator model.

10.7 Quark–Parton Phase Transitions

Let us go back to Fig. 10.4. The hadronic temperature T is plotted against $\ln(\tanh^2(\eta))$ or $\ln((v/c)^2)$. We can also plot $(v/c)^2$ as a function of T, as shown in Fig. 10.5.

As is seen in Fig. 10.5, the curve is nearly vertical for low temperature, but becomes nearly horizontal for high temperature, even though it is continuous. Thus, we are led to suspect a phase transition between these two different sections of the curve. Let us look at what happened inside the hadron.

If the hadron is at rest or its speed is very low, we use the quark model. If the hadron is moving with the speed close to that of light, we use the parton model. Since the constituents behave quite differently in these two models, we are confronted with the question of whether they can be described as two different limiting cases of one covariant entity.

This kind of question is not new in physics. Before 1905, Einstein faced the question of two different energy–momentum relations for massive and massless particles. He ended up with the formula $E = \sqrt{m^2 + p^2}$, which is widely known as his $E = mc^2$.

Our quark–parton puzzle is similar to that of energy–momentum relation, as illustrated in Table 1.2. The dynamics of the quark model is the same as that of the hydrogen atom, where two constituent particles are bound together by an attractive force.

We are all familiar with the dynamics of the hydrogen atom, but it is a challenge to understand Feynman's parton picture as a Lorentz-boosted hydrogen atom. The key variable is the time separation between the quarks, not seen in the Schrödinger picture of the hydrogen atom.

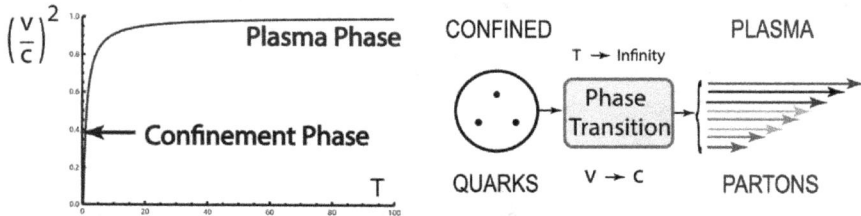

Fig. 10.5. Transition from the confinement phase to a plasma phase. The quarks are confined within a hadron when the hadron is at rest, but they behave like free particles when the hadron speed reaches that of light, and the temperature becomes very high. This figure is an extended interpretation of Fig. 10.4.

10.8 Wigner Functions and Uncertainty Relations

In the Wigner phase-space picture of quantum mechanics, both the position and momentum variables are c-numbers (Wigner, 1932). According to the uncertainty principle, we cannot determine the exact point in the phase space of the position and momentum variables, but we can localize these variables into an area in the phase space. The minimum area is of course Planck's constant. By studying the geometry of this area, we can study the uncertainty relation in more detail than in the Heisenberg or Schrödinger picture. This phase-space representation has been studied extensively in recent years in connection with quantum optics, and there are many books and review articles available on Wigner functions (Wigner, 1932; Feynman, 1998; Kim and Noz, 1991). Thus, it is not necessary to give a full-fledged introduction to this subject.

For two coordinate variables, the Wigner function is defined as (Kim and Noz, 1991)

$$W(x_1, x_2; p_1, p_2) = \left(\frac{1}{\pi}\right)^2 \int \exp\{-2i(p_1 y_1 + p_2 y_2)\}$$
$$\times \psi^*(x_1 + y_1, x_2 + y_2)\psi(x_1 - y_1, x_2 - y_2) \, dy_1 \, dy_2. \tag{10.46}$$

This function can therefore be derived from the density matrix:

$$W(x_1, x_2; p_1, p_2) = \left(\frac{1}{\pi}\right)^2 \int \exp\{-2i(p_1 y_1 + p_2 y_2)\}$$
$$\times \rho(x_1 - y_1, x_2 - y_2; x_1 + y_1, x_2 + y_2) \, dy_1 \, dy_2. \tag{10.47}$$

The Wigner function corresponding to the wave function of Eq. (10.10) is

$$W(x_1, x_2; p_1, p_2) = \left(\frac{1}{\pi}\right)^2 \exp\{-e^{2\eta}(x_1 - x_2)^2 - e^{-2\eta}(x_1 + x_2)^2$$
$$-e^{-\eta}(p_1 - p_2)^2 - e^{\eta}(p_1 + p_2)^2\}. \tag{10.48}$$

If we do not make observations of the $x_2 p_2$ coordinates and average over them, the Wigner function becomes (Yurke and Potasek, 1987; Ekert and Knight, 1989)

$$W(x_1, p_1) = \int W(x_1, x_2; p_1, p_2) \, dx_2 \, dp_2. \tag{10.49}$$

The evaluation of the integral leads to

$$W(x,p) = \left\{ \frac{1}{\pi^2(1 + \sinh^2 \eta)} \right\}^{1/2} \exp\left\{ -\left(\frac{x^2}{\cosh \eta} + \frac{p^2}{\cosh \eta} \right) \right\}, \quad (10.50)$$

where we have replaced x_1 and p_1 by x and p, respectively. This Wigner function gives an elliptic distribution in the phase space of x and p. This distribution gives the uncertainty product of

$$(\Delta x)^2 (\Delta p)^2 = \frac{1}{4}(1 + \sinh^2 \eta). \quad (10.51)$$

Because x_1 is coupled with x_2, our ignorance about the x_2 coordinate, which in this case acts as Feynman's rest of the universe, increases the uncertainty in the x_1 world which, in Feynman's words, is the system in which we are interested.

From this definition, it is straightforward to show (Kim and Noz, 1991)

$$Tr(\rho) = \int W(x,p) \, dx \, dp, \quad (10.52)$$

and

$$Tr(\rho^2) = 2\pi \int W^2(x,p) \, dx \, dp. \quad (10.53)$$

If we compute these integrals, $Tr(\rho) = 1$, as it should be for all pure and mixed states. On the other hand, the evaluation of the $Tr(\rho^2)$ integral leads to the result of Eq. (10.24).

What happens when we fail to observe the time-separation coordinate? We can approach this problem using the Wigner function (Wigner, 1932). For the density matrix of Eq. (10.28), the Wigner function is

$$W_\eta(z,t,p_z,p_0) = \frac{1}{\pi} \int \exp\left[(i(z'p_z + t'p_0)]\rho_\eta(z - z', t - t') \, dz' \, dt'. \quad (10.54)$$

The evaluation of this integral is straight-forward (Kim and Noz, 1991; Davies and Davies, 1975). For $\eta = 0$, the Wigner function becomes

$$W_0(z,p_z;t,p_0) = \left(\frac{1}{\pi}\right)^2 \exp\left[-(z^2 + p_z^2 + t^2 + p_0^2)\right]. \quad (10.55)$$

This is the Wigner function for the minimal uncertainty state without any thermal effects, or for the hadron at rest. This Gaussian form is illustrated in Fig. 10.6.

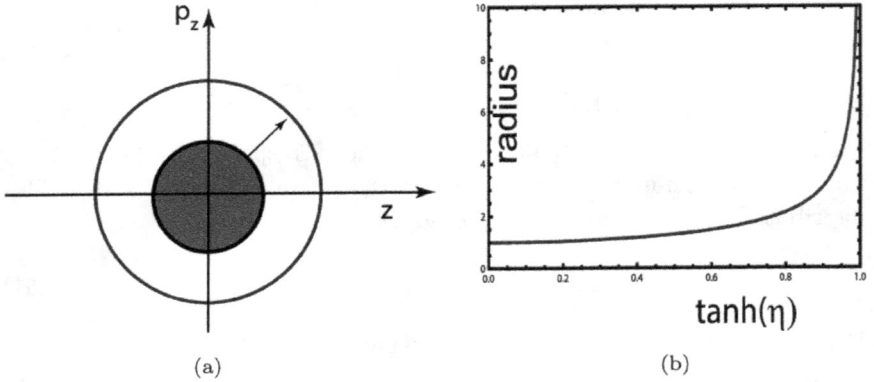

Fig. 10.6. Uncertainty distribution in the Wigner phase space. The radius takes the minimal value determined by the uncertainty principle when the hadron is at rest and its temperature is zero. The radius increases as the temperature rises as indicated in (a). The dependence of this radius on $\tanh(\eta)$ (the hadronic speed) is also given in (b).

For non-zero values of η, $W_\eta(z, p_z; t, p_0)$ becomes

$$\left(\frac{1}{\pi}\right)^2 \exp\left\{-\frac{1}{2}\left[e^{2\eta}(t+z)^2 + e^{-2\eta}(t-z)^2\right.\right.$$
$$\left.\left. + e^{-2\eta}(p_z - p_0)^2 + e^{2\eta}(p_z + p_0)^2\right]\right\}. \tag{10.56}$$

If we do not observe the second pair of variables, we have to integrate this function over t and p_0, and the resulting Wigner function is

$$W_\eta(z, p_z) = \int W(z, p_z; t, p_0)\, dt\, dp_0, \tag{10.57}$$

and the evaluation of this integration leads to (Han $et\ al.$, 1999)

$$W_\eta(x, p) = \frac{1}{\pi \cosh \eta} \exp\left[-\left(\frac{z^2 + p_z^2}{\cosh(2\eta)}\right)\right]. \tag{10.58}$$

The failure to make measurements on the time-separation variable leads to a radial expansion of the Wigner phase space as in the case of the thermal excitation. The radius is

$$\sqrt{\cosh(2\eta)} = \sqrt{\frac{1 + \tanh^2 \eta}{1 - \tanh^2 \eta}}. \tag{10.59}$$

As is indicated in Fig. 10.6, the radius becomes larger when $\tanh(\eta)$ becomes larger or the hadron moves with an increasing speed.

10.9 Lorentz-invariant Uncertainty Relation

The ground-state wave function for the covariant oscillator system takes the Gaussian form of

$$\psi_\eta(z,t) = \frac{1}{\sqrt{\pi}} \exp\left\{ -\frac{1}{4} \left[e^{-2\eta}(z+t)^2 + e^{2\eta}(z-t)^2 \right] \right\}. \qquad (10.60)$$

We can construct the momentum wave function by taking the Fourier transformation

$$\phi_\eta(p_z, p_0) = \frac{1}{2\pi} \int \psi_\eta(z,t) \exp\left\{ i\left(p_z z - p_0 t\right) \right\} dz\, dt \qquad (10.61)$$

which becomes

$$\frac{1}{\sqrt{\pi}} \exp\left\{ -\frac{1}{4} \left[e^{-2\eta}(p_z + p_0)^2 + e^{2\eta}(p_z - p_0)^2 \right] \right\}. \qquad (10.62)$$

Let us go back to Eq. (10.61), the exponent $(p_z z - p_0 t)$ is a Lorentz-invariant quantity and it can be written as

$$p_z z - p_0 t = \frac{1}{2}\left[(z+t)(p_z - p_0) + (z-t)(p_z + p_0) \right]. \qquad (10.63)$$

In this expression,

$$(z+t)(p_z - p_0) \quad \text{and} \quad (z-t)(p_z + p_0) \qquad (10.64)$$

are separately Lorentz-invariant. Thus, it is possible to define the uncertainty products along the light cones as shown in Fig. 10.7. Thus, the uncertainty products defined along the light cones remain invariant.

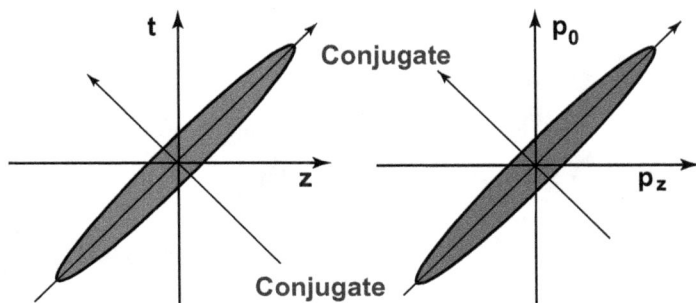

Fig. 10.7. Lorentz-invariant uncertainty products. The uncertainty products defined along the light cones remain invariant under the Lorentz boosts. This figure is derived from Fig. 10.2.

Chapter 11

Further Applications of the Lorentz Group

In an attempt to see whether Einstein's special relativity is applicable to localized waves for bound states, we had to develop the representations of the Lorentz group based on harmonic oscillators and 2×2 matrices. Since those oscillators and matrices serve as the basic scientific languages for many branches of physics, the formalism presented in this book could be applicable to some of those branches. Optical science is a case in point.

Optics was created much earlier than physics. Humans needed eye glasses, telescopes, and magnifiers many centuries before Newton wrote his first book on mechanics and optics. Optics books in those days had to be written without the mathematical tools available after Newton.

More recently, in 1937, Jenkins and White wrote their book entitled *Fundamental of Optics* (Jenkins and White, 1976), which still serves as an important reference book in optics. In this book, however, there are no matrices, no complex numbers, and no Fourier transformations. These days, it is not possible to write optics papers without harmonic oscillators and 2×2 matrices. In optics, 2×2 matrices are often called *ABCD* matrices.

In view of the progress made since the time of Jenkins and White, optics is now ready to make another jump in its scientific language: the jump to the Lorentz group. The authors of the present book published extensively on the Lorentz group in optics since 1988 (Han *et al.*, 1988; Kim and Noz, 1991; Başkal *et al.*, 2015). Let us see how the Lorentz group can serve as the basic scientific language for optical and information sciences.

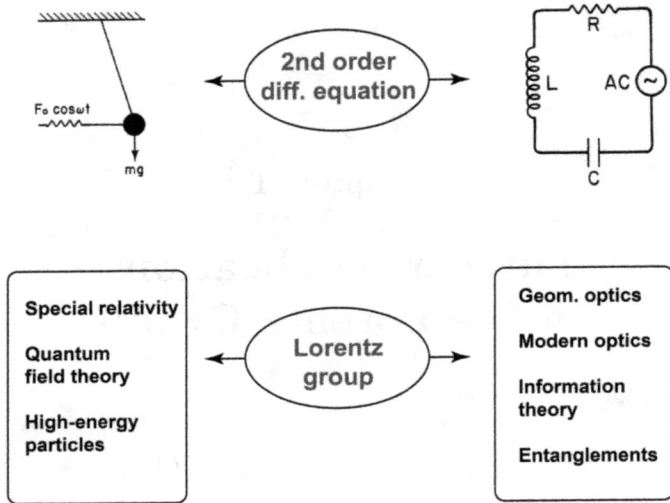

Fig. 11.1.   Bridge between high-energy physics and optics and information theory. The Lorentz group is the mathematical bridge between them.

11.1   Ray Optics

In ray optics, the optical ray moves along a straight line. The distance from the beam axis and its slope are the basic parameters. Thus, these two parameters can be grouped into one column matrix of the form

$$\begin{pmatrix} x \\ s \end{pmatrix}, \tag{11.1}$$

where x is for the distance from the axis of the ray, and s is the slope of the ray. For lenses and concave cavities, we are led to computations with matrices of the form

$$\begin{pmatrix} 1 & a \\ 0 & 1 \end{pmatrix} \quad \text{and} \quad \begin{pmatrix} 1 & 0 \\ b & 1 \end{pmatrix}. \tag{11.2}$$

These triangular matrices are of the form of Wigner's little group for massless particles, as noted in Chapter 6.

In multilayer optics, the optical ray is partially transmitted and partially reflected when it hits the surface of the material with a different index of refraction. This effect is treated by a 2×2 matrix corresponding to a Lorentz boost (Monzón and Sánchez-Soto, 1999, 2000; Georgieva and Kim, 2001).

The matrix algebra for these aspects of ray optics was systematically given in Chapter 6 of this book.

11.2 Polarization Optics

In polarization optics, we are dealing with two electric fields perpendicular to the direction of the propagation of light. If the beam is along the z direction, it can be written as

$$\begin{pmatrix} \psi_1 \\ \psi_2 \end{pmatrix} = \begin{pmatrix} a\, e^{-i\phi/2} \\ b\, e^{i\phi/2} \end{pmatrix} \exp\{-i(kz - \omega t)\}. \tag{11.3}$$

The column vector

$$\begin{pmatrix} a\, e^{-i\phi/2} \\ b\, e^{i\phi/2} \end{pmatrix} \tag{11.4}$$

is called the Jones vector (Jones, 1941, 1947). The parameters a and b are the amplitudes of the electric fields along the x and y directions respectively. The angle ϕ is the phase difference between the components.

The rotation around the z axis is trivial. The 2×2 matrix for this transformation is given in Chapter 6. The phase between the beams could also change. In addition, the amplitudes can undergo different rates of attenuation. All these are summarized in Table 11.1.

If the phase difference remains constant, the Jones vector contains enough information about the polarization of the beam. On the other hand,

Table 11.1 Polarization optics and special relativity sharing the same mathematics.

Polarization optics	Transformation matrix	Particle symmetry
Phase shift ϕ	$\begin{pmatrix} e^{-i\phi/2} & 0 \\ 0 & e^{i\phi/2} \end{pmatrix}$	Rotation around z
Rotation around z	$\begin{pmatrix} \cos(\theta/2) & -\sin(\theta/2) \\ \sin(\theta/2) & \cos(\theta/2) \end{pmatrix}$	Rotation around y
Squeeze along x and y	$\begin{pmatrix} e^{\mu/2} & 0 \\ 0 & e^{-\mu/2} \end{pmatrix}$	Boost along z
$\sin^2\alpha$	Determinant	$(\text{mass})^2$

Notes: Each matrix has its clear role in both optics and relativity. The determinant of the 2×2 matrix obtained from the Stokes vector or from the four-momentum remains invariant under Lorentz transformations. It is interesting to note that the decoherency parameter (least fundamental) in optics corresponds to the mass (most fundamental) in particle physics.

if the phase difference does not stay constant, we have to consider the degree of coherency. For this purpose, let us construct the 2×2 matrix:

$$C = \begin{pmatrix} S_{11} & S_{12} \\ S_{21} & S_{22} \end{pmatrix}, \tag{11.5}$$

with

$$\langle \psi_i^* \psi_j \rangle = \frac{1}{T} \int_0^T \psi_i^*(t + \tau) \psi_j(t) \, dt. \tag{11.6}$$

This 2×2 matrix is called the *coherency matrix* in the optics literature (Azzam and Bashara, 1999; Başkal *et al.*, 2015; Born and Wolf, 1999; Brosseau, 1998; Saleh and Teich, 2007). This is a convenient way of writing the four-Stokes parameters in the 2×2 form. The Stokes parameters constitute a four-component vector that transforms like the four-vector under Lorentz transformations (Han *et al.*, 1997b; Başkal *et al.*, 2015). Thus, coherency matrix is like the 2×2 matrix given in Eq. (6.1).

The four elements S_{ij} are the Stokes parameters (Azzam and Bashara, 1999; Hecht, 2002). They traditionally are grouped into one column matrix with four-elements. The 4×4 transformation matrices applicable to this column matrix are called Mueller matrices (Mueller, 1943; Hecht, 2002). It has been shown that the Mueller matrices are like those in the 4×4 representation of the Lorentz group (Han *et al.*, 1997a). It was also noted in Chapter 5 that the 4×4 algebra of the Lorentz group can be translated to that of 2×2 Lorentz group. The coherency matrix of Eq. (11.5) is like the 2×2 four-momentum matrix given in Eq. (6.1).

For the Jones vector given in Eq. (11.4), the C matrix can now be written as

$$C = \begin{pmatrix} a^2 & ab\, e^{+i\phi} \cos \alpha \\ ab\, e^{-i\phi} \cos \alpha & b^2 \end{pmatrix}, \tag{11.7}$$

where $\cos \alpha$ measures the degree of coherence. If the two beams are completely coherent, $\cos \alpha = 1$ with $\alpha = 0$. If the phase difference is completely random, $\cos \alpha = 0$.

The trace and determinant of the above coherency matrix are

$$\text{tr}(C) = a^2 + b^2,$$

$$\det(C) = (ab)^2 \sin^2 \alpha. \tag{11.8}$$

Let us go back to the matrix given in Eq. (6.1) for the four-momentum matrix. It is interesting to note that the determinant of that matrix is (mass)^2 of the particle. This corresponds to the degree of decoherence of the Stokes parameters as shown in Table 11.1.

The degree of polarization is defined as (Saleh and Teich, 2007)

$$f = \sqrt{1 - \frac{4 \det(C)}{(\mathrm{tr}(C))^2}} = \sqrt{1 - \frac{4(ab)^2 \sin^2 \alpha}{(a^2 + b^2)^2}}. \tag{11.9}$$

This degree is one if $\alpha = 0$. When $\alpha = 90^o$, it becomes

$$\frac{a^2 - b^2}{a^2 + b^2}. \tag{11.10}$$

Without loss of generality, we can assume that a is greater than b. If they are equal, this minimum degree of polarization is zero.

11.3 Coherent States and Squeezed States

The harmonic oscillator wave functions allow the following step-down and step-up operations.

$$\hat{a} \, \chi_n(x) = \sqrt{n} \, \chi_{n-1}(x) \quad \text{and} \quad \hat{a}^\dagger \, \chi_n(x) = \sqrt{n+1} \, \chi_{n+1}(x), \tag{11.11}$$

with

$$\hat{a} = \frac{1}{\sqrt{2}} \left(x + \frac{\partial}{\partial x} \right) \quad \text{and} \quad \hat{a}^\dagger = \frac{1}{\sqrt{2}} \left(x - \frac{\partial}{\partial x} \right). \tag{11.12}$$

In terms of these operators, the uncertainty relation can be written as

$$[\hat{a}, \hat{a}^\dagger] = 1. \tag{11.13}$$

We can now write the wave function $\chi_n(x)$ as a *ket vector* $|n\rangle$, and its Hermitian conjugate as $\langle n|$. The orthogonality relation can be written as

$$\langle n'|n \rangle = \delta_{n'n}. \tag{11.14}$$

These notations play the pivotal role in quantum field theory, where $\hat{a}^\dagger$ and $\hat{a}$ are used as the creation and annihilation operators applicable to the state $|n\rangle$ with n particles. For the world of photons, $|n\rangle$ is the state of n identical photons. Thus,

$$|n\rangle = \frac{1}{\sqrt{n!}} (\hat{a}^\dagger)^n |0\rangle, \quad \text{or} \quad (\hat{a}^\dagger)^n |0\rangle = \sqrt{n!} |n\rangle, \tag{11.15}$$

where $|0\rangle$ is for the zero-photon or vacuum state.

With these notations, we can now consider the state $|\alpha\rangle$ defined as

$$\exp\left\{i\left(\alpha\hat{a}^{\dagger} - \alpha^{*}\hat{a}\right)\right\}. \tag{11.16}$$

If this is applied to the zero-photon state:

$$|\alpha\rangle = \exp\left\{i\alpha\left(\hat{a}^{\dagger} - \hat{a}\right)\right\}|0\rangle, \tag{11.17}$$

where α is a complex number. The Taylor expansion of this expression becomes

$$|\alpha\rangle = \exp\left(-|\alpha|^2\right)/2\sum_{n}\frac{1}{\sqrt{n!}}|n\rangle. \tag{11.18}$$

This is known as the coherent state of light. Therefore, the physics of coherent state is the physics of harmonic oscillators.

It was noted in Sec. 10.9 that the Wigner function defined over the phase space is a convenient tool in studying uncertainty products. This aspect was explained in detail in a book entitled *Phase Space Picture of Quantum Mechanics* (Kim and Noz, 1991) written by the authors of the present book.

If there are two photons, whose numbers are created by $\hat{a}^{\dagger}$ and $\hat{b}^{\dagger}$ respectively, we can use the notation $|n, m\rangle$ where n and m are the a-type and b-type photon numbers respectively. We can consider the unitary operator:

$$B(\eta) = \exp\left\{-\frac{\eta}{2}\left(\hat{a}^{\dagger}\hat{b}^{\dagger} - \hat{a}\hat{b}\right)\right\}. \tag{11.19}$$

If this operator is applied to the vacuum state $|0, 0\rangle$, the result is

$$B(\eta)|0, 0\rangle = \frac{1}{\cosh\eta}\sum_{k}(\tanh\eta)^{k}|k, k\rangle. \tag{11.20}$$

This form is known as the squeezed vacuum (Yuen, 1976) or squeezed state. It is possible to produce this series using the covariant harmonic oscillators as given in Sec. 8.5. The operator $B(\eta)$ squeezes the vacuum wave functions. Thus, the series of Eq. (11.20) is called the squeezed state of the vacuum. This concept of squeeze comes from the Lorentz group as seen in Chapter 8. Thus, the squeezed state is the physics of the harmonic oscillators and 2×2 matrices.

The squeezed state of $|n, 0\rangle$ is given also in Sec. 8.5. As for the squeezed state of $|n, m\rangle$, the calculation is very complicated, but has been carried out by (Rotbart, 1981).

11.4 Entanglement Problems

From the mathematical point of view, the question of entanglement is whether the function separable in two variables becomes inseparable. Let us start with the Gaussian function of the form

$$\psi(x, y) = \frac{1}{\sqrt{\pi}} \exp\left\{ -\frac{1}{2} \left(x^2 + y^2 \right) \right\}. \tag{11.21}$$

This form is separable in the x and y variables. If this Gaussian function is squeezed to

$$\psi_\eta(x, y) = \frac{1}{\sqrt{\pi}} \exp\left\{ -\frac{1}{4} \left(e^{-2\eta}(x + y)^2 + e^{2\eta}(x - y)^2 \right) \right\}$$

$$= \frac{1}{\cosh \eta} \sum_k (\tanh \eta)^k \chi_k(x) \chi_k(y), \tag{11.22}$$

we say that this function is entangled (Giedke *et al.*, 2003; Kim and Noz, 2005; Walls and Milburn, 2008). This form of entanglement was discussed extensively in Chapter 8. We have discussed in detail how this form of entanglement leads to increase in entropy and temperature as shown in Secs. 10.5 and 10.6.

Let us next go back to the two-component system of polarized beam given in Sec. 11.2. If the two-beams do not interfere with each other, they are not entangled. If their phases become mixed up as given in Eq. (11.6), they become entangled, and the coherency matrix of Eq. (11.7) can be used as the density matrix if properly normalized. If we set $a = b = 1$, and $\phi = 0$ for simplicity, the density matrix becomes

$$\rho = \frac{1}{2} \begin{pmatrix} 1 & \cos \alpha \\ \cos \alpha & 1 \end{pmatrix}. \tag{11.23}$$

The trace of this matrix is one. This matrix is Hermitian and can therefore be diagonalized to

$$\frac{1}{2} \begin{pmatrix} 1 + \cos \alpha & 0 \\ 0 & 1 - \cos \alpha \end{pmatrix}. \tag{11.24}$$

This leads to the entropy of

$$S = -(1 + \cos \alpha) \ln \left(\frac{1 + \cos \alpha}{2} \right) - (1 - \cos \alpha) \ln \left(\frac{1 - \cos \alpha}{2} \right). \tag{11.25}$$

If the system is completely separate with $\cos \alpha = 1$, the entropy is zero with $S = 0$. If the system is thoroughly entangled with $\cos \alpha = 0$, the entropy becomes maximum with $S = 2 \ln(2)$.

Bibliography

Alkofer, R., Hll, A., Kloker, M., Krassnigg, A., and Roberts, C. D. (2005). On nucleon electromagnetic form factors, *Few-Body Systems* **37**, 1–2, pp. 1–31, doi:10.1007/s00601-005-0110-6, http://link.springer.com/10.1007/s00601-005-0110-6.

Anderson, C. D. (1933). The Positive Electron, *Physical Review* **43**, 6, pp. 491–494, doi:10.1103/PhysRev.43.491, https://link.aps.org/doi/10.1103/PhysRev.43.491.

Applebaum, A. (2017). *Between East and West: Across the Borderlands of Europe* (Random House Publishing company, New York, NY, USA), ISBN 978-0-525-43318-7, Originally published 1994; OCLC: 962008162.

Azzam, R. M. A.-G. and Bashara, N. M. (1999). *Ellipsometry and Polarized Light*, 4th Edn., North-Holland personal library (Elsevier, Amsterdam), ISBN 978-0-444-87016-2, Originally published 1977; OCLC: 247501433.

Bargmann, V. (1947). Irreducible unitary representations of the Lorentz group, *The Annals of Mathematics* **48**, 3, p. 568, doi:10.2307/1969129, http://www.jstor.org/stable/1969129?origin=crossref.

Başkal, S., Kim, Y. S., and Noz, M. E. (2014). Wigner's space-time symmetries based on the two-by-two matrices of the damped harmonic oscillators and the Poincaré sphere, *Symmetry* **6**, 3, pp. 473–515, doi:10.3390/sym6030473, http://www.mdpi.com/2073-8994/6/3/473/.

Başkal, S., Kim, Y. S., and Noz, M. E. (2015). *Physics of the Lorentz Group* (IOP Publishing), ISBN 978-1-68174-254-0, http://iopscience.iop.org/book/978-1-6817-4254-0, doi:10.1088/978-1-6817-4254-0.

Başkal, S., Kim, Y. S., and Noz, M. E. (2016). Entangled harmonic oscillators and space-time entanglement, *Symmetry* **8**, 7, pp. 55–80, doi:10.3390/sym8070055, http://www.mdpi.com/2073-8994/8/7/55.

Başkal, S., Kim, Y. S., and Noz, M. E. (2017). Loop representation of Wigner's little groups, *Symmetry* **9**, 7, pp. 97–118, doi:10.3390/sym9070097, http://www.mdpi.com/2073-8994/9/7/97.

Bég, M. A. B., Lee, B. W., and Pais, A. (1964). SU(6) and electromagnetic interactions, *Physical Review Letters* **13**, 16, pp. 514–517, doi:10.1103/PhysRevLett.13.514, https://link.aps.org/doi/10.1103/PhysRevLett.13.514.

Bell, J. S. (2004). *Speakable and Unspeakable in Quantum Mechanics: Collected Papers on Quantum Philosophy*, Revised Edn. (Cambridge University Press, Cambridge, UK; New York, NY USA), ISBN 978-0-521-81862-9, present edition: (Bell and Aspect, 2008).

Bell, J. S. and Aspect, A. (2008). *Speakable and Unspeakable in Quantum Mechanics*, 2nd Edn., Collected Papers on Quantum Philosophy (Cambridge Univ. Press, Cambridge, UK), ISBN 978-0-521-52338-7 978-0-521-81862-9, includes bibliographical references; Previous ed.: 1987; OCLC: 552116191.

Bjorken, J. D. and Paschos, E. A. (1969). Inelastic electron–proton and γ-proton scattering and the structure of the nucleon, *Physical Review* **185**, 5, pp. 1975–1982, doi:10.1103/PhysRev.185.1975, https://link.aps.org/doi/10.1103/PhysRev.185.1975.

Blum, K. (2012). *Density Matrix Theory and Applications*, 3rd Edn., No. 64 in Springer Series on Atomic, Optical and Plasma Physics (Springer, Heidelberg; New York), ISBN 978-3-642-20560-6, Originally published in 1981.

Born, M. (1938). A suggestion for unifying quantum theory and relativity, *Proceedings of the Royal Society A: Mathematical, Physical and Engineering Sciences* **165**, 921, pp. 291–303, doi:10.1098/rspa.1938.0060, http://rspa.royalsocietypublishing.org/cgi/doi/10.1098/rspa.1938.0060.

Born, M. (1949). Reciprocity theory of elementary particles, *Reviews of Modern Physics* **21**, 3, pp. 463–473, doi:10.1103/RevModPhys.21.463, https://link.aps.org/doi/10.1103/RevModPhys.21.463.

Born, M. and Wolf, E. (1999). *Principles of Optics: Electromagnetic Theory of Propagation, Interference and Diffraction of Light*, 7th Edn. (Cambridge University Press, Cambridge; New York), ISBN 978-0-521-64222-4 978-0-521-63921-7, Originally published 1980 by Pergamon Press, Oxford, UK.

Brosseau, C. (1998). *Fundamentals of Polarized Light: A Statistical Optics Approach* (Wiley, New York), ISBN 978-0-471-14302-4, "A Wiley-Interscience publication.

Buras, A. J. (1980). Asymptotic freedom in deep inelastic processes in the leading order and beyond, *Reviews of Modern Physics* **52**, 1, pp. 199–276, doi:10.1103/RevModPhys.52.199, https://link.aps.org/doi/10.1103/RevModPhys.52.199.

Capri, A. Z. and Chiang, C. C. (1976). Extended meson fields. An alternative to quark confinement, *Il Nuovo Cimento A* **36**, 4, pp. 331–353, doi:10.1007/BF02724520, http://link.springer.com/10.1007/BF02724520.

Capri, A. Z. and Chiang, C. C. (1977). Extended baryon fields and baryon–meson interaction, *Il Nuovo Cimento A* **38**, 2, pp. 191–208, doi:10.1007/BF02724541, http://link.springer.com/10.1007/BF02724541.

Condon, E. U. and Shortley, G. H. (1979). *The Theory of Atomic Spectra*, reprinted Edn. (Univ. Pr, Cambridge), ISBN 978-0-521-09209-8, Originally published 1951; OCLC: 258057861.

Davies, R. and Davies, K. (1975). On the Wigner distribution function for an oscillator, *Annals of Physics* **89**, 2, pp. 261–273, doi:10.1016/0003-4916(75) 90182-7, http://linkinghub.elsevier.com/retrieve/pii/0003491675901827.

Dirac, P. A. M. (1927). The quantum theory of the emission and absorption of radiation, *Proceedings of the Royal Society A: Mathematical, Physical and Engineering Sciences* **114**, 767, pp. 243–265, doi: 10.1098/rspa.1927.0039, http://rspa.royalsocietypublishing.org/cgi/doi/10. 1098/rspa.1927.0039.

Dirac, P. A. M. (1945a). Application of quaternions to lorentz transformations, *Proceedings of the Royal Irish Academy. Section A: Mathematical and Physical Sciences* **50**, 1944/1945, pp. 261–270, http://www.jstor.org/stable/ 20520646.

Dirac, P. A. M. (1945b). Unitary representations of the Lorentz group, *Proceedings of the Royal Society A: Mathematical, Physical and Engineering Sciences* **183**, 994, pp. 284–295, doi:10.1098/rspa.1945.0003, http://rspa. royalsocietypublishing.org/cgi/doi/10.1098/rspa.1945.0003.

Dirac, P. A. M. (1949). Forms of relativistic dynamics, *Reviews of Modern Physics* **21**, 3, pp. 392–399, doi:10.1103/RevModPhys.21.392, https://link.aps.org/ doi/10.1103/RevModPhys.21.392.

Dirac, P. A. M. (1963). A remarkable representation of the 3 + 2 de Sitter group, *Journal of Mathematical Physics* **4**, 7, pp. 901–909, doi:10.1063/1.1704016, http://aip.scitation.org/doi/10.1063/1.1704016.

Einstein, A., Podolsky, B., and Rosen, N. (1935). Can quantum-mechanical description of physical reality be considered complete? *Physical Review* **47**, 10, pp. 777–780, doi:10.1103/PhysRev.47.777, https://link.aps.org/doi/10. 1103/PhysRev.47.777.

Ekert, A. K. and Knight, P. L. (1989). Correlations and squeezing of two-mode oscillations, *American Journal of Physics* **57**, 8, pp. 692–697, doi:10.1119/ 1.15922, http://aapt.scitation.org/doi/10.1119/1.15922.

Fano, U. (1957). Description of states in quantum mechanics by density matrix and operator techniques, *Reviews of Modern Physics* **29**, 1, pp. 74–93, doi:10.1103/RevModPhys.29.74, https://link.aps.org/doi/10. 1103/RevModPhys.29.74.

Fetter, A. L. and Walecka, J. D. (2003). *Quantum Theory of Many-Particle Systems* (Dover Publications, Mineola, NY), ISBN 978-0-486-42827-7, Originally published: San Francisco, McGraw-Hill, c1971.

Feynman, R. (1969a). The behavior of hadron collisions at extreme energies, in C. Yang *et al.* (Eds.), *Proceedings of the 3rd International Conference on High Energy Collisions* (Gordon and Breach, New York, NY, USA), pp. 237–249, Stony Brook, New York, USA, 5-6-September.

Feynman, R. P. (1969b). Very high-energy collisions of hadrons, *Physical Review Letters* **23**, 24, pp. 1415–1417, doi:10.1103/PhysRevLett.23.1415, https:// link.aps.org/doi/10.1103/PhysRevLett.23.1415.

Feynman, R. P. (1998). *Statistical Mechanics: A Set of Lectures,* Advanced Book Classics (Westview Press, Boulder, Colo), ISBN 978-0-201-36076-9, Originally published 1972; OCLC: ocm60679997.

Feynman, R. P., Kislinger, M., and Ravndal, F. (1971). Current matrix elements from a relativistic quark model, *Physical Review D* **3**, 11, pp. 2706–2732, doi:10.1103/PhysRevD.3.2706, https://link.aps.org/doi/10.1103/PhysRevD.3.2706.

Frazer, W. R. and Fulco, J. R. (1960). Effect of a pion–pion scattering resonance on nucleon structure. II, *Physical Review* **117**, 6, pp. 1609–1614, doi:10.1103/PhysRev.117.1609, https://link.aps.org/doi/10.1103/PhysRev.117.1609.

Fujimura, K., Kobayashi, T., and Namiki, M. (1970). Nucleon electromagnetic form factors at high momentum transfers in an extended particle model based on the quark model, *Progress of Theoretical Physics* **43**, 1, pp. 73–79, doi:10.1143/PTP.43.73, https://academic.oup.com/ptp/article-lookup/doi/10.1143/PTP.43.73.

Gell-Mann, M. (1964). A schematic model of baryons and mesons, *Physics Letters* **8**, 3, pp. 214–215, doi:10.1016/S0031-9163(64)92001-3, http://linkinghub.elsevier.com/retrieve/pii/S0031916364920013.

Georgieva, E. and Kim, Y. S. (2001). Iwasawa effects in multilayer optics, *Physical Review E* **64**, 2, doi:10.1103/PhysRevE.64.026602, https://link.aps.org/doi/10.1103/PhysRevE.64.026602.

Giedke, G., Wolf, M. M., Krger, O., Werner, R. F., and Cirac, J. I. (2003). Entanglement of formation for symmetric Gaussian states, *Physical Review Letters* **91**, 10, doi:10.1103/PhysRevLett.91.107901, https://link.aps.org/doi/10.1103/PhysRevLett.91.107901.

Ginzburg, V. and Man'ko, V. (1965). Relativistic oscillator models of elementary particles, *Nuclear Physics* **74**, 3, pp. 577–588, doi:10.1016/0029-5582(65)90203-8, http://linkinghub.elsevier.com/retrieve/pii/0029558265902038.

Guillemin, V. and Sternberg, S. (2001). *Symplectic Techniques in Physics,* Reprinted Edn. (Cambridge Univ. Press, Cambridge), ISBN 978-0-521-38990-7, Originally published 1984; OCLC: 248721606.

Hamermesh, M. (1989). *Group Theory and Its Application to Physical Problems,* Dover Books on Physics and Chemistry (Dover Publications, New York), ISBN 978-0-486-66181-0, Originally published by Addison-Wesley, Reading, MA 1962.

Han, D. and Kim, Y. S. (1980). Yukawa's approach and Dirac's approach to relativistic quantum mechanics: Relativistic harmonic oscillator model, *Progress of Theoretical Physics* **64**, 5, pp. 1852–1860, doi:10.1143/PTP.64.1852, https://academic.oup.com/ptp/article-lookup/doi/10.1143/PTP.64.1852.

Han, D. and Kim, Y. S. (1981). Little group for photons and gauge transformations, *American Journal of Physics* **49**, 4, pp. 348–351, doi:10.1119/1.12509, http://aapt.scitation.org/doi/10.1119/1.12509.

Han, D. and Kim, Y. S. (1988). Special relativity and interferometers, *Physical Review A* **37**, 11, pp. 4494–4496, doi:10.1103/PhysRevA.37.4494, https://link.aps.org/doi/10.1103/PhysRevA.37.4494.

Han, D., Kim, Y. S., and Noz, M. E. (1981). Physical principles in quantum field theory and in covariant harmonic oscillator formalism, *Foundations of Physics* **11**, 11–12, pp. 895–905, doi:10.1007/BF00727106, http://link.springer.com/10.1007/BF00727106.

Han, D., Kim, Y. S., and Noz, M. E. (1988). Linear canonical transformations of coherent and squeezed states in the Wigner phase space, *Physical Review A* **37**, 3, pp. 807–814, doi:10.1103/PhysRevA.37.807, https://link.aps.org/doi/10.1103/PhysRevA.37.807.

Han, D., Kim, Y. S., and Noz, M. E. (1997a). Jones-matrix formalism as a representation of the Lorentz group, *Journal of the Optical Society of America A* **14**, 9, p. 2290, doi:10.1364/JOSAA.14.002290, https://www.osapublishing.org/abstract.cfm?URI=josaa-14-9-2290.

Han, D., Kim, Y. S., and Noz, M. E. (1997b). Stokes parameters as a Minkowskian four-vector, *Physical Review E* **56**, 5, pp. 6065–6076, doi:10.1103/PhysRevE.56.6065, https://link.aps.org/doi/10.1103/PhysRevE.56.6065.

Han, D., Kim, Y. S., and Noz, M. E. (1999). Illustrative example of Feynman's rest of the universe, *American Journal of Physics* **67**, 1, pp. 61–66, doi:10.1119/1.19192, http://aapt.scitation.org/doi/10.1119/1.19192.

Han, D., Kim, Y. S., Noz, M. E., and Yeh, L. (1993). Symmetries of two-mode squeezed states, *Journal of Mathematical Physics* **34**, 12, pp. 5493–5508, doi:10.1063/1.530318, http://aip.scitation.org/doi/10.1063/1.530318.

Han, D., Kim, Y. S., and Son, D. (1982). E(2)-like little group for massless particles and neutrino polarization as a consequence of gauge invariance, *Physical Review D* **26**, 12, pp. 3717–3725, doi:10.1103/PhysRevD.26.3717, https://link.aps.org/doi/10.1103/PhysRevD.26.3717.

Han, D., Kim, Y. S., and Son, D. (1983). Gauge transformations as Lorentz-Boosted rotations, *Physics Letters B* **131**, 4-6, pp. 327–329, doi:10.1016/0370-2693(83)90509-9, http://linkinghub.elsevier.com/retrieve/pii/0370269383905099.

Han, D., Kim, Y. S., and Son, D. (1986a). Eulerian parametrization of Wigner's little groups and gauge transformations in terms of rotations in two-component spinors, *Journal of Mathematical Physics* **27**, 9, pp. 2228–2235, doi:10.1063/1.526994, http://aip.scitation.org/doi/10.1063/1.526994.

Han, D., Kim, Y. S., and Son, D. (1986b). Photons, neutrinos, and gauge transformations, *American Journal of Physics* **54**, 9, pp. 818–821, doi:10.1119/1.14454, http://aapt.scitation.org/doi/10.1119/1.14454.

Hecht, E. (2002). *Optics*, 4th Edn. (Addison-Wesley, Reading, Mass), ISBN 978-0-8053-8566-3.

Heisenberg, W. (1989). Encounters and Conversations with Albert Einstein, in *ENCOUNTERS WITH EINSTEIN and Other Essays on People, Places, and Particles* (Princeton University Press, Princeton, N.J), ISBN 978-0-691-02433-2, pp. 107–122.

Heitler, W. (1984). *The Quantum Theory of Radiation*, 3rd Edn. (Dover Publications, New York), ISBN 978-0-486-64558-2.

Hofstadter, R. (1956). Electron scattering and nuclear structure, *Reviews of Modern Physics* **28**, 3, pp. 214–254, doi:10.1103/RevModPhys.28.214, https://link.aps.org/doi/10.1103/RevModPhys.28.214.

Hofstadter, R. and McAllister, R. W. (1955). Electron scattering from the proton, *Physical Review* **98**, 1, pp. 217–218, doi:10.1103/PhysRev.98.217, https://link.aps.org/doi/10.1103/PhysRev.98.217.

Howard, D. A. (2005). Albert Einstein as a philosopher of science, *Physics Today* **58**, 12, doi:10.1063/1.2169442, http://physicstoday.scitation.org/doi/10.1063/1.2169442.

Hussar, P. E. (1981). Valons and harmonic oscillators, *Physical Review D* **23**, 11, pp. 2781–2783, doi:10.1103/PhysRevD.23.2781, https://link.aps.org/doi/10.1103/PhysRevD.23.2781.

Hussar, P. E., Kim, Y. S., and Noz, M. E. (1980). Three-particle symmetry classifications according to the method of Dirac, *American Journal of Physics* **48**, 12, pp. 1038–1042, doi:10.1119/1.12301, http://aapt.scitation.org/doi/10.1119/1.12301.

Hwa, R. C. (1980). Evidence for valence-quark clusters in nucleon structure functions, *Physical Review D* **22**, 3, pp. 759–764, doi:10.1103/PhysRevD.22.759, https://link.aps.org/doi/10.1103/PhysRevD.22.759.

Hwa, R. C. and Zahir, M. S. (1981). Parton and valon distributions in the nucleon, *Physical Review D* **23**, 11, pp. 2539–2553, doi:10.1103/PhysRevD.23.2539, https://link.aps.org/doi/10.1103/PhysRevD.23.2539.

Inönü, E. and Wigner, E. P. (1953). On the contraction of groups and their representations, *Proceedings of the National Academy of Sciences* **39**, 6, pp. 510–524, doi:10.1073/pnas.39.6.510, http://www.pnas.org/cgi/doi/10.1073/pnas.39.6.510.

Iwasawa, K. (1949). On some types of topological groups, *The Annals of Mathematics* **50**, 3, p. 507, doi:10.2307/1969548, http://www.jstor.org/stable/1969548?origin=crossref.

Janner, A. and Janssen, T. (1971). Electromagnetic compensating gauge transformations, *Physica* **53**, 1, pp. 1–27, doi:10.1016/0031-8914(71)90098-X, http://linkinghub.elsevier.com/retrieve/pii/003189147190098X.

Jenkins, F. A. and White, H. E. (1976). *Fundamentals of Optics*, 4th Edn. (McGraw-Hill, New York), ISBN 978-0-07-032330-8, first ed. published in 1937 under title: Fundamentals of physical optics, includes index.

Jones, R. C. (1941). A new calculus for the treatment of optical systems. Description and discussion of the calculus, *Journal of the Optical Society of America* **31**, 7, p. 488, doi:10.1364/JOSA.31.000488, https://www.osapublishing.org/abstract.cfm?URI=josa-31-7-488.

Jones, R. C. (1947). A new calculus for the treatment of optical systems V. A more general formulation, and description of another calculus, *Journal of the Optical Society of America* **37**, 2, p. 107, doi:10.1364/JOSA.37.000107, https://www.osapublishing.org/abstract.cfm?URI=josa-37-2-107.

Karr, T. (1976). *Field Theory of Extended Hadrons Based on Covariant Harmonic Oscillators*, Ph.D. Thesis, University of Maryland, College Park, Maryland.

Kim, Y. S. (1976). Model for relativistic bound-state perturbation theory, *Physical Review D* **14**, 1, pp. 273–279, doi:10.1103/PhysRevD.14.273, https://link.aps.org/doi/10.1103/PhysRevD.14.273.

Kim, Y. S. (1989). Observable gauge transformations in the parton picture, *Physical Review Letters* **63**, 4, pp. 348–351, doi:10.1103/PhysRevLett.63.348, https://link.aps.org/doi/10.1103/PhysRevLett.63.348.

Kim, Y. S. (1998). Does Lorentz boost destroy coherence? *Fortschritte der Physik* **46**, 6-8, pp. 713–723, doi:10.1002/(SICI)1521-3978(199811)46:6/8⟨713:: AID-PROP713⟩3.0.CO;2-H, http://doi.wiley.com/10.1002/%28SICI%2915 21-3978%28199811%2946%3A6/8%3C713%3A%3AAID-PROP713%3E3. 0.CO%3B2-H.

Kim, Y. S. and Noz, M. E. (1973). Covariant harmonic oscillators and the quark model, *Physical Review D* **8**, 10, pp. 3521–3527, doi:10.1103/PhysRevD.8.3521, https://link.aps.org/doi/10.1103/PhysRevD.8.3521.

Kim, Y. S. and Noz, M. E. (1977). Covariant harmonic oscillators and the parton picture, *Physical Review D* **15**, 1, pp. 335–338, doi:10.1103/PhysRevD.15.335, https://link.aps.org/doi/10.1103/PhysRevD.15.335.

Kim, Y. S. and Noz, M. E. (1986). *Theory and Applications of the Poincaré Group* (Springer Netherlands, Dordrecht), ISBN 978-94-010-8526-7 978-94-009-4558-6, http://link.springer.com/10.1007/978-94-009-4558-6, doi: 10.1007/978-94-009-4558-6.

Kim, Y. S. and Noz, M. E. (1991). *Phase Space Picture of Quantum Mechanics: Group Theoretical Approach*, no. v. 40 in Lecture notes in physics series (World Scientific, Singapore; Teaneck, NJ), ISBN 978-981-02-0360-3 978-981-02-0361-0.

Kim, Y. S. and Noz, M. E. (2005). Coupled oscillators, entangled oscillators, and Lorentz-covariant harmonic oscillators, *Journal of Optics B: Quantum and Semiclassical Optics* **7**, 12, pp. S458–S467, doi:10.1088/1464-4266/7/12/005, http://stacks.iop.org/1464-4266/7/i=12/a=005?key=crossref.8970fab 58458b312ba68e38d4ed3c24a

Kim, Y. S. and Noz, M. E. (2011). Lorentz harmonics, squeeze harmonics and their physical applications, *Symmetry* **3**, 4, pp. 16–36, doi:10.3390/sym3010016, http://www.mdpi.com/2073-8994/3/1/16/.

Kim, Y. S., Noz, M. E., and Oh, S. H. (1979a). Representations of the Poincaré group for relativistic extended hadrons, *Journal of Mathematical Physics* **20**, 7, pp. 1341–1344, doi:10.1063/1.524237, http://aip.scitation.org/doi/10.1063/1.524237, see also, Physics Auxiliary Publication Service Document No. PAPS JMAPA-20-1336-12.

Kim, Y. S., Noz, M. E., and Oh, S. H. (1979b). A simple method for illustrating the difference between the homogeneous and inhomogeneous Lorentz groups, *American Journal of Physics* **47**, 10, pp. 892–897, doi: 10.1119/1.11622, http://aapt.scitation.org/doi/10.1119/1.11622.

Kim, Y. S. and Wigner, E. P. (1987). Cylindrical group and massless particles, *Journal of Mathematical Physics* **28**, 5, pp. 1175–1179, doi:10.1063/1. 527824, http://aip.scitation.org/doi/10.1063/1.527824.

Kim, Y. S. and Wigner, E. P. (1990a). Entropy and Lorentz transformations, *Physics Letters A* **147**, 7, pp. 343–347, doi:10.1016/0375-9601(90)90550-8, http://linkinghub.elsevier.com/retrieve/pii/0375960190905508.

Kim, Y. S. and Wigner, E. P. (1990b). Space–time geometry of relativistic particles, *Journal of Mathematical Physics* **31**, 1, pp. 55–60, doi:10.1063/1. 528827, http://aip.scitation.org/doi/10.1063/1.528827.

Kupersztych, J. (1976). Is there a link between gauge invariance, relativistic invariance and electron spin? *Il Nuovo Cimento B Series 11* **31**, 1, pp. 1–11, doi:10.1007/BF02730313, http://link.springer.com/10.1007/ BF02730313.

Licht, A. L. and Pagnamenta, A. (1970). Wave functions and form factors for relativistic composite particles. I, *Physical Review D* **2**, 6, pp. 1150–1156, doi: 10.1103/PhysRevD.2.1150, https://link.aps.org/doi/10.1103/PhysRevD.2. 1150.

Magnus, W., Oberhettinger, F., and Soni, R. P. (1966). *Formulas and Theorems for the Special Functions of Mathematical Physics* (Springer-Verlag, Berlin; New York), ISBN 978-3-662-11761-3 978-3-662-11763-7, http:// books.google.com/books?id=KoJQAAAAMAAJ, oCLC: 557712575.

Markov, M. (1956). On dynamically deformable form factors in the theory of elementary particles, *Il Nuovo Cimento* **3**, S4, pp. 760–772, doi:10.1007/ BF02746074, http://link.springer.com/10.1007/BF02746074.

Matevosyan, H. H., Thomas, A. W., and Miller, G. A. (2005). Study of lattice QCD form factors using the extended Gari–Krmpelmann model, *Physical Review C* **72**, 6, doi:10.1103/PhysRevC.72.065204, https://link.aps.org/ doi/10.1103/PhysRevC.72.065204.

Mita, K. (1978). Euler–Lagrange equation and conservation laws for bilocal fields, *Physical Review D* **18**, 12, pp. 4545–4547, doi:10.1103/PhysRevD.18.4545, https://link.aps.org/doi/10.1103/PhysRevD.18.4545.

Monzón, J. J. and Sánchez-Soto, L. L. (1999). Fully relativisticlike formulation of multilayer optics, *Journal of the Optical Society of America A* **16**, 8, p. 2013, doi:10.1364/JOSAA.16.002013, https://www.osapublishing.org/ abstract.cfm?URI=josaa-16-8-2013.

Monzón, J. J. and Sánchez-Soto, L. L. (2000). Fresnel formulas as Lorentz transformations, *Journal of the Optical Society of America A* **17**, 8, p. 1475, doi:10.1364/JOSAA.17.001475, https://www.osapublishing.org/ abstract.cfm?URI=josaa-17-8-1475.

Mueller, H. (1943). Memorandum on the polarization optics of the photoelastic shutter, Tech. Rep. Report No. 2 of the OSRD project, OEMsr-576. Office of Scientific Research and Development of the United States.

Naimark, M. (1954). Linear representation of the Lorentz group, *Usp. Mat. Nauk* **9**, pp. 19–93, Naimark M A 1957 Linear representation of the Lorentz group *Am. Math. Soc. Transl. Ser. 2 6 379–458* (Engl. transl.) Naimark M A

1964 Linear Representations of the Lorentz Group (International Series of Monographs in Pure and Applied Mathematics vol 63) (Oxford: Pergamon) (Engl. transl.).

Ojima, I. (1981). Gauge fields at finite temperatures–"Thermo field dynamics" and the KMS condition and their extension to gauge theories, *Annals of Physics* **137**, 1, pp. 1–32, doi:10.1016/0003-4916(81)90058-0, http://linkinghub.elsevier.com/retrieve/pii/0003491681900580.

Oz-Vogt, J., Mann, A., and Revzen, M. (1991). Thermal coherent states and thermal squeezed states, *Journal of Modern Optics* **38**, 12, pp. 2339–2347, doi:10.1080/09500349114552501, http://www.tandfonline.com/doi/abs/10.1080/09500349114552501.

Planck, M. (1900). Zur Theorie des Gesetzes der Energieverteilung im Normalspectrum, *Verhandlungen der Deutschen Physikalischen Gesellschaft* **2**, pp. S.237–245, translated into English by D. ter Haar and S. G. Brush, published in Planck's Original Papers in Quantum Physics edited by H. Kangrao (Taylor and Francis, London, 1972).

Rotbart, F. C. (1981). Complete orthogonality relations for the covariant harmonic oscillator, *Physical Review D* **23**, 12, pp. 3078–3080, doi:10.1103/PhysRevD.23.3078, https://link.aps.org/doi/10.1103/PhysRevD.23.3078.

Ruiz, M. J. (1974). Orthogonality relation for covariant harmonic-oscillator wave functions, *Physical Review D* **10**, 12, pp. 4306–4307, doi:10.1103/PhysRevD.10.4306, https://link.aps.org/doi/10.1103/PhysRevD.10.4306.

Saleh, B. E. A. and Teich, M. C. (2007). *Fundamentals of Photonics*, 2nd Edn., Wiley series in pure and applied optics (Wiley Interscience, Hoboken, N.J), ISBN 978-0-471-35832-9.

Schultz, S. (2004). Newly discovered diary chronicles Einstein's last years, *Princeton Weekly Bulletin* **93**, 25.

Sogami, I. (1973). Reconstruction of non-local field theory. I: Causal description, *Progress of Theoretical Physics* **50**, 5, pp. 1729–1747, doi:10.1143/PTP.50.1729, https://academic.oup.com/ptp/article-lookup/doi/10.1143/PTP.50.1729.

Tanikawa, Y. (1979). *Hideki Yukawa. Scientific Works* (Iwanami Shoten, Tokyo, Japan).

Umezawa, H., Matsumoto, H., and Tachiki, M. (1982). *Thermo Field Dynamics and Condensed States* (North-Holland Pub. Co.; Sole distributors for the U.S.A. and Canada, Elsevier Science Pub. Co, Amsterdam; New York, N.Y), ISBN 978-0-444-86361-4.

von Neumann (1932). *Die Mathematische Grundlagen der Quanten-mechanik* (Springer, Berlin, Germany).

Von Neumann, J. (1996). *Mathematical Foundations of Quantum Mechanics*, Princeton landmarks in mathematics and physics (Princeton University Press, Princeton, NJ USA), ISBN 978-0-691-02893-4 978-0-691-08003-1, Originally published 1955; OCLC: ocm37904902.

Walls, D. F. and Milburn, G. J. (2008). *Quantum Optics*, 2nd Edn. (Springer, Berlin), ISBN 978-3-540-28573-1.

Weinberg, S. (1964a). Feynman rules for any spin, *Physical Review* **133**, 5B, pp. B1318–B1332, doi:10.1103/PhysRev.133.B1318, https://link.aps.org/doi/10.1103/PhysRev.133.B1318.

Weinberg, S. (1964b). Feynman rules for any spin. II. Massless particles, *Physical Review* **134**, 4B, pp. B882–B896, doi:10.1103/PhysRev.134.B882, https://link.aps.org/doi/10.1103/PhysRev.134.B882.

Weinberg, S. (1964c). Photons and gravitons in S-matrix theory: Derivation of charge conservation and equality of gravitational and inertial mass, *Physical Review* **135**, 4B, pp. B1049–B1056, doi:10.1103/PhysRev.135.B1049, https://link.aps.org/doi/10.1103/PhysRev.135.B1049.

Weinberg, S. (1966). Dynamics at infinite momentum, *Physical Review* **150**, 4, pp. 1313–1318, doi:10.1103/PhysRev.150.1313, https://link.aps.org/doi/10.1103/PhysRev.150.1313.

Wigner, E. P. (1931). *Gruppentheorie und ihre Anwendung auf die Quantenmechanik der Atomspektren* (Friedrich Vieweg und Sohn, Braunsweig, Germany).

Wigner, E. P. (1932). On the quantum correction for thermodynamic equilibrium, *Physical Review* **40**, 5, pp. 749–759, doi:10.1103/PhysRev.40.749, https://link.aps.org/doi/10.1103/PhysRev.40.749.

Wigner, E. P. (1939). On unitary representations of the inhomogeneous Lorentz group, *The Annals of Mathematics* **40**, 1, pp. 149–204, doi:10.2307/1968551, http://www.jstor.org/stable/1968551?origin=crossref.

Wigner, E. P. (1959). *Group Theory and its Applications to the Quantum Theory of Atomic Spectra* (Academic Press, New York, NY USA), translated from the German by J. J. Griffin.

Wigner, E. P. (1960a). Normal form of antiunitary operators, *Journal of Mathematical Physics* **1**, 5, pp. 409–413, doi:10.1063/1.1703672, http://aip.scitation.org/doi/10.1063/1.1703672.

Wigner, E. P. (1960b). Phenomenological distinction between unitary and antiunitary symmetry operators, *Journal of Mathematical Physics* **1**, 5, pp. 414–416, doi:10.1063/1.1703673, http://aip.scitation.org/doi/10.1063/1.1703673.

Wigner, E. P. and Yanase, M. M. (1963). Information contents of distributions, *Proc. Nat. Acad. Scie* **49**, 6, pp. 910–918.

Yuen, H. P. (1976). Two-photon coherent states of the radiation field, *Physical Review A* **13**, 6, pp. 2226–2243, doi:10.1103/PhysRevA.13.2226, https://link.aps.org/doi/10.1103/PhysRevA.13.2226.

Yukawa, H. (1935). On the interaction of elementary particles, I. *Proc. Phys. Math. Japan* **17**, pp. 48–57.

Yukawa, H. (1949). On the radius of the elementary particle, *Physical Review* **76**, 2, pp. 300–301, doi:10.1103/PhysRev.76.300.2, https://link.aps.org/doi/10.1103/PhysRev.76.300.2.

Yukawa, H. (1953). Structure and mass spectrum of elementary particles. I. General considerations, *Physical Review* **91**, 2, pp. 415–416, doi:10.1103/PhysRev.91.415.2, https://link.aps.org/doi/10.1103/PhysRev.91.415.2.

Yurke, B., McCall, S. L., and Klauder, J. R. (1986). SU(2) and SU(1,1) interferometers, *Physical Review A* **33**, 6, pp. 4033–4054, doi:10.1103/ PhysRevA.33.4033, https://link.aps.org/doi/10.1103/PhysRevA.33.4033.

Yurke, B. and Potasek, M. (1987). Obtainment of thermal noise from a pure quantum state, *Physical Review A* **36**, 7, pp. 3464–3466, doi:10.1103/ PhysRevA.36.3464, https://link.aps.org/doi/10.1103/PhysRevA.36.3464.

Index

www.ingramcontent.com/pod-product-compliance
Lightning Source LLC
Chambersburg PA
CBHW050603190326
41458CB00007B/2160